INTERDISCIPLINA
APPROACHES TO
TEACHING ART IN

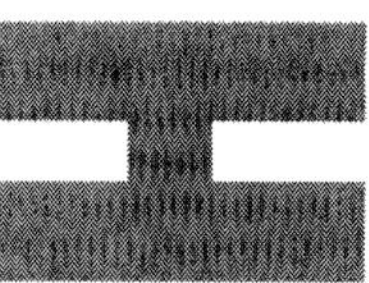

Pamela G. Taylor
Virginia Commonwealth University, Richmond

B. Stephen Carpenter, II
Texas A & M University, College Station

Christine Ballengee-Morris
The Ohio State University, Columbus

Billie Sessions
California State University, San Bernardino

2006

The National Art Education Association

About NAEA

The National Art Education Association is the world's largest professional visual arts education association and a leader in educational research, policy, and practice for art education. NAEA's mission is to advance art education through professional development, service, advancement of knowledge, and leadership.

Membership includes elementary and secondary art teachers, middle school and high school students in the National Art Honor Society programs, artists, administrators, museum educators, arts council staff, university professors, and students from the United States and several foreign countries. It also includes publishers, manufacturers, and suppliers of art materials; parents; students; retired art educators; and others concerned about quality art education in our schools.

NAEA publishes *Art Education*, *Studies in Art Education*, and other professional papers on art education; holds an annual convention; conducts research; sponsors a teacher awards program; develops standards for student learning, school programs, and teacher preparation; and co-sponsors workshops, seminars, and institutes on art education. For further information, visit our website at www.arteducators.org.

1806 Robert Fulton Drive
Reston, VA 20191

To order a copy of this book or obtain additional information, contact National Art Education Association: www.arteducators.org/store or 800-299-8321.

Order No. 226
ISBN 1-890160-35-0

Acknowledgements

RECOGNITION AND APPRECIATION

We would like to thank the many high school art teachers who answered our call and submitted wonderful, exciting stories of interdisciplinary plans and experiences—Tara Adams, Sara Brock, Mary Chin, Philip Conrad, Sara Gant, Dr. Elisa Gargarella, Casie Geiswite, Ann Gerald, Anna Golden, Donna Green, Kathleen Hall, Chris Kitzmiller, Claudia Michael, Dr. Mark Moilanen, Judy O'Neal Miller, Jodi Patterson, Stacy Potter, Kathleen Ragusea, Sue Raymond, Dan Springer, Denise Stanley, Colette Stemple, Alicia Swackhamer, Jennifer Trettner, Anna Ursyn, Don Wass, Shari Williams, and Lance Wurst.

We also thank our own high school art teachers who revealed first-hand how glorious the study and making of art in high school can be. We extend our sincere thanks to Virginia Commonwealth University graduate student, Kathryn Helms, who tirelessly read the book many times and assisted greatly in editing and layout design. We are also extremely grateful for Donna Green's willingness to share her professional photography expertise and produce the exciting images we used in this book as visual metaphors. We thank Juliet Tzou, a graduate student at Texas A & M University for her assistance.

We are honored and humbled by the willingness of our professors, mentors, and good friends Dr. Brent Wilson and Dr. Marjorie Wilson to contribute the foreword to this book and to Dr. Howard Risatti for sharing his insights in the afterword.

We acknowledge the kindnesses of artists Joseph Norman, Kendall Buster, Robert Irwin, James Luna, Jeremy Mitzel, Brenda Stein, Donna Rizzo, and Claudia Lee for sharing images of their exceptional work in this book. We also acknowledge the J. Paul Getty Trust, Scala/Art Resource, Octagon Earthworks, and TETAC (Transforming Education through the

Arts Challenge) for print permissions.

We thank our spouses and significant others—both human and canine—Ron, Zoubeida, David, Inky and Brooke for their willingness to deal with our sleepless nights, blurred visions, and constant excuses. Thanks, too is owed to our families for understandinging and supporting the time we needed to dedicate to this project. We offer special appreciation to our students, our respective universities, administrators, and colleagues for their support.

We would be remiss if we did not thank the National Art Education Association for enabling the publication of this book. We also owe a debt of gratitude to our professors and friends, past and present, at the Pennsylvania State University for their role in our individual and shared foundation, courage, and confidence to undertake this endeavor— Marjorie Wilson, Brent Wilson, Charles Garoian, Paul Bolin, Elizabeth Garber, Yvonne Gaudelius, Patricia Amburgy, Al Anderson, the late Robert Ott, Jeanne Brady, Bonnie MacDonald, the late Tim Gallucci, Shirley Yokley, Angela La Porte, John White, Peg Speirs, Tim Jackson, Leisha Jones, David Gall, Melinda Mayer, Paul Briggs, Steve Darnell, Ruth Slotnick, Lydia Dambekalns, Cheryl Capezzuti, Travis DiNicola, and James Ritchey. The intense spirit of collaboration, encouragement, camaraderie, and friendships we experienced during our respective graduate studies at Penn State provoked our life-long professional and personal relationship.

We are, by all accounts, four art teachers who share a similar passion and dedication to the field of art education, as do other teachers of visual art. That said, we want to acknowledge all art teachers around the world who are driven by a boundless energy, creative spirit, and unwavering commitment to art education.

And finally, we appreciate and welcome you, the reader who is now in a position to share in the process of making this book meaningful. May this book offer you countless opportunities to inquire about the world, make connections among various ways of knowing, and expand what it means to teach and learn in, through, with, and because of art.

-Pam, Steve, Christine, and Billie

TABLE OF CONTENTS

Foreword

By Wilson and Wilson

For those of you who know Dr. Marjorie Wilson and Dr. Brent Wilson — both of whom recently retired from the Art Education Program faculty at the Pennsylvania State University—their following approach to the Foreword of this book will not surprise you. For those of you who do not know them, theirs was and is an enduring personal and professional relationship that thrives on contention, debate, and the fervently honest collaborative spirit with which we approached the writing of this book. We therefore, present this Foreword in two parts. The first is Brent Wilson's statement followed by a response from Marjorie Wilson. Enjoy!

Brent Wilson: "I believe someone . . . who is a writer is not simply doing his work in his books, but that his major work is, in the end, himself in the process of writing his books"—Foucault[1]

If, indeed, a work—a text— is the writer, him or herself, and if, as further expounded in Foucault's *Archeology of Knowledge,* "the work includes the whole life as well as the text" (Ryan, 1993, p. 14), then what makes *Interdisciplinary Approaches to Teaching Art in High School* truly exciting is that it is a text which embodies the lives and knowledge and works of four separate and unique writers combined seamlessly into a single work. What, then, is the text? When each of those writers is an experienced high school art teacher and a university professor of art education, when those authors are experienced researchers and theoreticians, and when those authors are also knowing inhabitants of a surprisingly large number of domains within the worlds of art, folk arts and crafts, digital hypertextual realms, and visual culture, what happens is something extraordinary. Were we to clone the combination of Pamela G. Taylor, B. Stephen Carpenter, II, Christine Ballengee-Morris, and Billie Sessions, we would have something akin to a composite-super-self—an amazingly knowledgeable, experienced, practiced, and superbly synthesized high school art teacher, the likes of which we wish we would find in every high school art classroom. Barring this, however, the text, *Interdisciplinary Approaches* to *Teaching Art in High School,* provides the sources, knowledge, skills, understanding, insight, and the vision that we would wish every high school teacher, beginning or experienced, to know and practice.

Let us point to just a few of the dimensions of the lives, experiences, expertise, research, pedagogy, and artistry from which this text emerges.

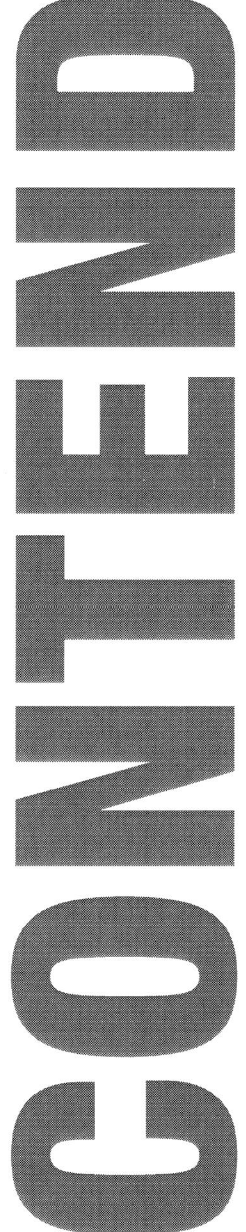

What should art education in the contemporary high school look like? We should expect that a new art educational text would exemplify the latest thinking about new media and new art forms—including electronic and hypertextual ways of building knowledge and creating artworks. Pamela G. Taylor has a prophetic vision of how electronic technologies and especially hypertexts can revolutionize art education—no, not just art education, all education. Wouldn't it be marvelous if we were able to look into a human's mind to see a few million of the myriad bits of knowledge lodged there in the form of beliefs, thoughts, images, impressions, feelings, assumptions, purposes, values, and on and on? Think of our being able to view how one human connects countless bits of information one to another and another. As an art teacher, think what it would be like to see the network of connection as a student relates one remembered image to another and another and another in the process of creating and interpreting artworks. Think what it would be like for an entire classroom of students to create a record of all the images and texts that they are able to relate to an artwork they decide to study. If we were to see things such as this, we would begin to realize what enlightened art education is. And through the hypertext webs that Pamela Taylor, her colleagues, and her students have already constructed we can see these things.

But we also have every reason to expect that a new text in art education would be just as sensitive to the qualities of charcoal and paint. Wouldn't our ideal teacher be as expert with, say, clay, glazes, and ceramic forms as with electronic media? We've just described—in part—B. Stephen Carpenter, II, an accomplished exhibiting ceramic artist who is stretching the boundaries of his art form, and venturing into territories of installation and intermedia, a ceramic artist who is as comfortable with the language of ceramics criticism as he is with slabs of clay and the potter's wheel. Indeed he has earned the respect of ceramic artists for his research into the languages of ceramics interpretation and criticism. He shows us the ideal teacher who is both a maker of forms and one who can help students reveal the meanings of works of art, craft and visual culture.

And wouldn't we wish for our ideal teacher of art to understand cultural contexts—to know of the conditions, the norms, the ideas, and the social contexts that surround art objects? That teacher is one like Christine Ballengee-Morris, who knows realms of art, visual and material culture, folk art and craft from the vantage point of an insider and an insightful outsider. She knows what is authentic and what is not, she is a trusted guide who understands the complex issues which surround cultural artifacts—their creation, authenticity, ownership, possession, preservation, duplication, misrepresentation, interpretation, interrelationships, and transformations through time. In American schools where cultural identities of students are both constructed and refined, wouldn't we wish to have a teacher who has keen insights into the ways visual cultural items and artifacts shape and reshape changing identities and contribute to diversity and community—a guide who sees issues from their practical and theoretical dimensions?

And while we are constructing our composite super-art-teacher—a composite of our four authors—we might wish for an experienced high school teacher who can take a collection of visual cultural artifacts, say ceramic works, and from them construct a curriculum based on a visual world history that reveals the relationships among tableware and the decorative, ceramic objects that echo sociological and ideological issues, works that connect through their cultural roots, contain ceremony and are gods to their makers. Through works made of clay, Billie Sessions shows the range of our human passions, interests, desires, habits, and much, much more, all revealed through a monumental theory of world pottery and object making. And while we are commenting on her work, we should note that here is a teacher who has used her insight into the lives of ceramic educators to become a noted curator preparing the catalogue and organizing the coast-to-coast traveling exhibition of Bauhaus ceramist Marguerite Wildenhain, whose work and teaching was celebrated next to many of the students she influenced. Yes, we would wish for our teacher to be an insightful historian, sensitive interpreter, and visual cultural biographer.

Altered books are inventive examples of re-writing the text. Pictured here are Katie Helms's alteration of D. H. Lawrence's *Women in Love* (left) and Scott McCarney's[4] *DIDEROT/DOUBLEDAY/DECONSTRUCTION VOL VI* (1995). According to McCarney, "The image of a bookbinder is revealed by chiseling away pages of a discarded 20th century encyclopedia. The act of carving a pattern of 'pixels' into the book has re-organized the information it contains. 200 years after Diderot composed his own encyclopedia, modern technology has provided us with alternative means for sifting through data. The presentation of information becomes a part of its meaning. The question arises, 'is the book itself a technology of the past?' Only time will tell." Permission to print from Virginia Commonwealth University James Branch Cabell Library Special Collections (http://www.library.vcu.edu/jbc/speccoll/bookart.html). (See Figure I.1 on page 4.)

I remember the time during the 90s when these four gifted authors, researchers, teachers, and artists were at Penn State, as a golden era. Of course, Penn State has had other golden ages during the Lowenfeld[2] era and the Beittel[3] era. The 90s was the golden era for Wilson and Wilson. It was these four authors and their colleagues that made it so. Pamela Taylor once said, "I evaluate my students, not on how much they have learned from me, but on how much I have learned from them." What we have learned from our students is incalculable, and now the world of art education has the opportunity to learn from them as well.

Marjorie Wilson: That is all well and good— from a strictly Modernist or Prestructuralist point of view. And you have justly given credit to our fabulous foursome, but as for what you—and Foucault—want to call the text... it has been more than twenty years since Roland Barthes made the important distinction between the Work and the Text. What you describe is the Work. In Heartney's *Postmodernism,* she makes the point:

> The Work takes us back to the prestructural realm, where there is a stable external world from which an artwork or piece of writing issues. The reader's job is simply to interpret, or in Barthes' term, 'consume' it, in accordance with the creator's intentions. The Text, by contrast, is composed of that web of interwoven signifiers and deferred meanings that is poststructuralism. Or as Barthes describes it, the Text is 'a multi dimensional space in which a variety of writings, none of them original, blend and clash. The text is a tissue of quotations….'

You go on to say that the "text" should be consumed, that through following "the sources, knowledge, skills, understanding, insight, and the vision" provided, we would have the ideal high school art teacher.

Au contraire, that is precisely what the text argues against. If you read carefully, you will find innumerable urgings in *Interdisciplinary Approaches to Teaching Art in High School* to make notations, to add to, to find other pathways, to go in directions unexpected and unknown. In fact, the ideal format for this book would be as an electronic hypertext, where the invitation for response and argument could be easily and, more importantly, immediately, carried out. For, as Heartney[5] goes on to say:

> The Text has no author, or at least, no privileged figure who can be lifted above any of the raw materials out of which it is composed. In the end, the Text is created, not by the author, but by the reader who engages with it and puts it to work. Barthes uses the metaphor

of music to make his point, likening the Text to a score, which the performer brings to life.

To all those who interact as reader/writers of *Interdisciplinary Approaches to Teaching Art in High School,* as Tevya sings, in "Fiddler on the Roof," L'chaim! To life!!!!

Notes

[1] Ryan, A. (1993). *Foucault's life and hard times.* New York Review of Books, XL(7), 12-17.

[2]The Pennsylvania State University became an international center for art education when Austrian-born Viktor Lowenfeld joined the faculty in 1946. Lowenfeld was considered the most influential art educator of the 20th century and wrote one of the field's dominant books, *Creative and Mental Growth*, based on his pioneering work in psychology and the art of the visually impaired. (http://www.imakenews.com/psaanews/e_article000541989.cfm?x=b11,0,w)

[3] Often considered uncoventional and always highly active, the reknown potter and Penn State art educator Kenneth R. Beittel was influential in beginning *Studies in Art Education* and instrumental in founding the Seminar for Research in Art Education. After graduating from Penn State he was hired in 1953 by Lowenfeld, wrote the celebrated *Zen and the Art of Pottery*, and had more than twenty-five solo exhibitions in the United States and Japan. However, his memoirs suggest that his greatest achievement was his work with graduate students at Penn State. (http://www.sva.psu.edu/arted/program/historical/beittel.htm)

[4] Related sites on the artist, Scott McCarney
http://www.ithacafinechocolates.com/scottmccarney.html
http://www.pabagallery.com/artists.cfm?task=detail&id=38

[5] Heartney, E. (2001). Introduction. In *Postmodernism* (pp. 6-12). Cambridge, UK: Cambridge University Press

Dr. Brent Wilson is Emeritus Professor of Art Education at Penn State. Dr. Wilson's research includes studies of children's artistic development, the influences of popular visual culture on children's drawings, children's graphic narratives, Japanese and Taiwanese teenagers, dojinshi/manga, and children's interpretations of artworks. Wilson has authored or coauthored six books, 27 book chapters, 90 monographs and evaluative reports to agencies, and 60 articles in refereed journals. With Marjorie Wilson, he received the Manuel Barkan Award for their outstanding contribution to the literature of art education. He also received the Edwin Ziegfeld Award for contributions to international art education and research and the Lowenfeld Prize for his studies of children's art. In 1998, Dr. Wilson conducted research into school arts education programs for the President's Committee on the Arts and the Humanities. From 1988 to 1996 he evaluated Getty Education Institute professional development programs and published his findings in a book titled The Quiet Evolution: Changing the Face of Arts Education (1997). In 1987, Dr. Wilson served as a consultant to the National Endowment for the Arts where he drafted Towards Civilization, a report to the President and Congress on the status of arts education in the United States. From 1967 to 1982, he served as principal consultant to the National Assessment of Educational Progress in Art.

Dr. Marjorie Wilson retired as associate professor of art education at the Pennsylvania State University in 2005. Her research centers on the pedagogical applications of hypermedia and technology in art education. Most recently, she has pursued her profound interest in contemporary art forms and has developed and taught courses in performance, installation and video and their relationship to teaching. She has published more than thirty articles in professional journals and books and has co-authored three books. She has presented papers relating to her research throughout the United States and Canada, in The Netherlands, in Copenhagen, Denmark, and, most recently, in Norway where she shared her work relating to the confluence of performance and teaching. Dr. Marjorie Wilson was a researcher for the J. Paul Getty Trust in Los Angeles, California, and the author of a series of guides for teachers, published and distributed with print reproductions of major works of art, pointing to the cultural, historical, philosophical and critical aspects of those works. With Dr. Brent Wilson, she was awarded the Educational Press Association's Distinguished Achievement Award, given for excellence in educational journalism for a series of three articles published in School Arts, and the Manuel Barkan Award for outstanding contribution to the literature of art education.

Introduction

PURPOSES AND POSSIBILITIES

The idea for this book began many years ago during our high school art teaching experiences. Like many other high school art teachers, we worked very hard to make sure our students had opportunities to study and create meaningful and technically proficient works of art in a variety of media. We lobbied for funding, negotiated time and space, and promoted our programs through exhibitions, advanced placement connections, and simple yet spirited lunchroom or teachers' lounge discussions.

Unlike many teachers of the other disciplines, we embraced the integration of block scheduling that changed our daily schedules from six or seven fifty-minute class periods to four ninety-minute periods. Biased in our belief that the study and making of art was the only discipline that promoted inventive thinking, we felt that we simply had a longer time to do what we were already doing. That is, we worked with students in much the same way our college professors worked with us in our undergraduate and graduate studio classes—practicing techniques and making things called art. With the added time we felt that our students would become proficient with techniques and media earlier in the semester, allowing for the teaching and learning of more complex approaches before the end of the term. Much to our dismay, we soon discovered that this was not to be the case. Our high school students were not at the same interest, maturity, or focus levels as those of the college art majors in the studio courses we sought to emulate. Our high school students were easily distracted. They simply could not be truly engaged in one mode of working for a ninety-minute segment of time. We learned that employing a variety of instructional strategies was essential for making effective use of the time we had with our students. We worked to involve criticism activities that went beyond the studio critique. We conducted in-process group and self-critique discussions

Intentions and Invitations — Connections
Structure and Design — Guide — Contributions

Link Connect Mediate Relate Correlate
Associate Interact Rally Cooperate
Correspond Assemble Combine

and assigned writing tasks. Collaborative activities encouraged the students to work together to explore problems as well as make communally meaningful works of art. Aesthetic games challenged them to think and ponder in significant as well as inventive ways. Using portions of films and videos helped the students focus on a particular idea or mode of working. Student-initiated inquiry and research revealed to them that the study and making of art extended beyond the boundaries of our art rooms. We took this idea further by moving our classes to other areas of the school to discover different perspectives or ways of seeing. We conducted portions of our class under the tables as a way to experience multiple perspectives and lead philosophical debates inspired by the variety of brand names and designer labels that adorned the clothing and accessories of student apparel.

In our process of facilitating and developing these and other multiple strategies, we realized how extensive the study of art and artmaking must be in order for it to be meaningful to the lives of our students. Whether in the science class, the English classroom or the lunch cafeteria, students and teachers are surrounded by visual images. The process of creating, using and interpreting these visual images is unique to the study of art in some ways, but universal and applicable to a variety of disciplines in other ways.

Noticing, interpreting, and formulating inventive solutions and codes for communicating are integral to the study of math, science, and English. Social studies students work to recognize that visual images and artifacts represent cultural modes of representation as do foreign language classes. Applying and understanding contextual implications of historical depictions encourages a discerning approach to the study of history. Drama and music students perform through a connective and personifying interpretation of characters, roles, and musical association. Most vocational classes encourage observation and inventive problem solving. Observation skills regarding physical space and time are an integral focus of physical education classes.

Like many other art teachers, throughout our connective musings, we avoided labeling what we did. As a defense mechanism, we shied away from using terms like "interdisciplinary" to describe our instructional approaches. We taught art, not science or math, and it was the art-specific standards of learning and curriculum that were our primary concern. Our readings and observations of interdisciplinary approaches using art in other subjects concerned us also because of the "handmaiden" manner in which non-art teachers routinely incorporated art activities into their lessons. Our concern stemmed too from what we considered trivial interdisciplinary artmaking requests—advising a social studies teacher with a bulletin board display, solving structural problems and designing the set for the school play, or doing the layout for science fair flyers – that were more about materials and illustration and less about authentic art learning.

Such interdisciplinary concerns could cause many of us to comfortably dismiss these ideas and retreat further back to our art rooms where the study of art is central and where real world issues don't matter. But, we know better than that. We know that the study and making of art is essential to our students. Simply, we want the best for them. We also know that although art education has intrinsic qualities and learning opportunities that are idiosyncratic to the art class, there are many art and artmaking skills and ways of knowing that can be catalysts for learning across the disciplinary boundaries in high school.

Recognizing these possibilities is the first step. Leading rather than following or acquiescing to directives is the next. High school art teachers are dedicated intellectuals who know the value of learning, growth, and change. They recognize that challenges are opportunities for creativity and inventiveness. Granted, the challenges of incorporating interdisciplinary teaching and learning in high school art classes are many—varied grade levels and classes of students, standardized test-driven curriculum in other academic subjects, multiple artistic interest levels of students and teachers, time, class offerings, materials, extracurricular and co-curricular activities. But, challenges and problems are the roots of our creative explorations as artists and educators. Artists and educators thrive on their creative desire to solve problems and invent new ways of seeing, knowing, thinking, and learning.

Through the writing of this book, we were compelled to offer opportunities for interdisciplinary inspiration, with an emphasis on high school art teachers' creative drives and passionate beliefs in the power of art to educate. Like the study and making of art, we recognize the importance of creative freedom in teaching and learning art, especially at the high school level. Therefore, this book provides multiple ideas outlined in such a way that diverse approaches are encouraged and provoked.

Begun as a fairly simple desire to illustrate interdisciplinary teaching and learning possibilities in high school art, the process with which we approached, grappled, and finally wrote this book took us on a complicated yet connected journey of reflective discovery and possibility. Because we are such good friends, we related easily our teaching stories. However, correlating our belief systems was rarely an easy task. At times, we marveled at the association of four such differently-minded individuals and yet, each and every time our interactions became passionate and fervid discussions, we came away learning and understanding so much more about our field, ourselves, and each other. We rallied in our cooperative spirit realizing that so many aspects of high school art teaching (from classroom management to computers and the school cafeteria to paints and brushes) affect what as well as how we teach and learn. We honestly yet thoughtfully mediated our criticism. We corresponded carefully and sometimes silently as we realized the impossibility of addressing the vast implications of such an undertaking. And we assembled this book through a combined effort and dedication to our steadfast belief in the power of art and art education. In other words, we lived and continue to be affected by that which we have experienced through the process of writing this book. Interdisciplinary and/or integrated learning is indeed a mindset to which we have gladly relinquished our lives and relationships. It is our hope that this book will assist high school art teachers and future teachers to lead the way toward an assertive form of interdisciplinary instruction that will enhance their lives as well.

We structured and designed this book as both a guide and a provocative device for teachers and students of high school art education theory and practice. That is, we took a comprehensive approach and included a number of issues, topics, theories, and practices that, though specific to high school art education, may not be typically associated with interdisciplinary theory. We contend that teaching and learning in the visual arts is fundamentally interdisciplinary and integrated. Therefore, much of what we do as high school art teachers is intricately connected to other realms of knowledge and experience. And much of what our high school students do and think about affect the way both they and we make these connections. For example, many high school art teachers would agree that classroom management is crucial to their effectiveness as teachers and their students' ability to learn. Important too is recognizing and indeed valuing the individual interests, needs, and maturity of high school students. Just as understanding cultural diversity and respect for values other than our own are important life-long sensibilities, they too, play an important part in our ability to cooperate and make associations. Ideas, concepts, and essential questions provoke a more expansive and critical relationship with learning. Interdisciplinary ways of knowing may alter our aesthetics or views of what art is or should be as well as our instructional strategies, our views of staff development, and the ways we advocate or promote our art programs. To be sure, the fundamental interdisciplinary nature of how we live as human beings in a world linked and mediated by technology and visual culture is the driving force behind this book.

We used linking verbs to organize the chapters—connect, relate, corre-

Figure I.1. We invite you, the reader to comment, argue, suggest, and make links throughout this book in whatever manner you choose. Photograph by Donna Green, 2006.

late, associate, interact, rally, cooperate, mediate, correspond, assemble, and combine. Chapter topics include current developments in interdisciplinary theory, using the high school curriculum structure to an interdisciplinary arts advantage, critical thinking and cultural connections, student-initiated integrative learning, standards, assessment, and interactive computer technology, along with specific models for interdisciplinary approaches to service-learning and the study of visual culture. We use gray text boxes at the beginning of each chapter to identify specific issues, models and ways of working addressed in the chapter. Throughout the book, we include stories of the exemplary work that many high school art teachers graciously shared with us **(using parentheses to identify the stories as teacher contributions)**. We have also used stories from our own high school teaching experiences to illustrate specific interdisciplinary points, theories, and/or arguments as well as some fictional blends of our collective high school teaching memories. As we feel strongly that art education teachers and students should be knowledgeable of multiple viewpoints, the discussion questions and activities at the end of each chapter challenge the reader to move beyond this text through the exploration of other readings and research.

Although the chapters may build upon each other, by no means does a reader have to approach this book in a linear fashion. We have attempted to provide contextual clues and links within the chapters in the form of parenthetical connections, text pullouts, images, suggested readings, and endnotes. Originally, we planned to create even more connections between and among the information in this book through design devices such as lines, arrows, multiple text boxes, and notes. But, just as we recognized that we could not include all of the vast implications and connections for interdisciplinary approaches to high school art education in this one book, neither was it possible to visually make all of the connections. Therefore, we chose to approach the many interdisciplinary ways of teaching, learning and knowing as well as the connections between them provocatively. In other words, we hope that our words, propositions, links to other information, and suggested questions and activities will inspire the reader to continually pursue other ideas, voices, and ways of working throughout their teaching and learning as well as in the reading of this book. And therefore, we invite you the reader to comment, argue, suggest, and make links throughout this book in whatever manner that works best for you. You may for example, prefer to use a highlighter, pen or pencil to make marks directly on the text. You may wish to write in the margins or use the front and back papers to make notes, or stick notes or bookmarks in the book itself. (See Figure I.1). Our grandest wish is to see copies of this book tattered, torn, and marked all over as such use is the greatest confirmation of what compelled us to write this book in the first place—to simply provoke all of us to think, teach, and learn in more personally and artfully connective ways.

Chapter 1

ART AS A COHESIVE AGENT OF INTERDISCIPLINARITY

Brooke[1] is a typical high school junior. She rides to school every morning with her friends and arrives flushed and eager to see everyone in the hallways and at the lockers before classes begin for the day. Once a week, the first bell of the day directs Brooke to homeroom, where she and other juniors typically complete forms and other school paperwork as well as learn about specific opportunities and requirements for her grade level. From there she goes to her first class, Algebra III, with other juniors. In this class, Brooke and her friends engage in learning theorems and solving math problems. The walls in the math classroom feature posters of important mathematicians' work such as Euclid and Pythagoras. Rulers, calculators, protractors, books, and documents fill the worktables and shelves as papers constantly shuffle, notebook pages turn, and the teacher checks for understanding through sample and collaborative problem solving. After 90 minutes, the bell rings and Brooke and her friends dash through the crowded hallways to their lockers or other meeting places from which they make their way to their second period French II class.

In French class, a mixed-class of sophomores, juniors, and seniors read and translate stories, engage in dialogue with their teacher and each other, and respond to quiz questions on vocabulary and grammar. A map of France graces one wall, while posters of the Eiffel Tower and other French tourist hot spots, such as the gardens and vistas of Arles and Giverny, and the cathedral in Strasbourg fill another.

Advocacy — Careers — Art is an Essential Life Skill

Models and Approaches — Perspective and Culture

By third period, Brooke anxiously awaits lunch both as a time to eat and as a longer period in which to talk with her friends. Brooke eats during what is referred to as "second lunch" and therefore must attend 1/2 of her third period class, world history before and after lunch. In this class, Brooke, a few other juniors, and some sophomores and seniors listen as the teacher lectures, shows images, and writes important dates and historical events on the board, overhead projector, or projected computer screen. World maps, posters of important world leaders, and a time-line fill the walls. Shelves contain encyclopedias, textbooks, and other historical reference materials.

At lunch, Brooke stands in line and chats with friends while making meal choices from sandwiches, meats, vegetables, salad, milk, and ice cream. At their lunch table Brooke and her friends eat quickly. They laugh, point, and plan for after school event; they also complain about family, teachers, homework, and boredom. Posters and flyers pertaining to clubs, sports, events, and other co- and extracurricular events in school fill the cafeteria walls. A few posters about nutrition and some framed student artwork are also on display. Several teachers and administrators meander through the cafeteria. They talk with each other and with students as they maintain a presence and watch for disruptive behavior.

Brooke's fourth period class is Art I where she works alongside freshmen, sophomores, and seniors as well as a few fellow juniors. In this class, Brooke follows her teacher's instructions and experiments with different materials. She and her classmates view and discuss images of works of art they see in books and on a projection screen. Art posters, student artwork, classroom rules, and media-specific procedures fill the walls. Shelves and cabinets are full of supplies and materials and each student has a cubby or tray to keep their supplies and art works in-progress. The art teacher typically walks around the room and helps individual students with their work or busily gives out materials, cleans artmaking tools and paint containers, works on bulletin boards, or chats with student groups. Brooke and her friends often impatiently listen to the teacher's instructions for the day as they anxiously wait to continue their lunchtime discussion while working on their art. Ten minutes before the last bell of the day, Brooke quickly cleans her area, stores her work-in-progress, gathers her bags, and anticipates the upcoming search for other students in the school's crowded and noisy hallways. Once together, Brooke and her friends quickly resume their mid-day conversation and share their latest news while they ready themselves to drive home, attend after school jobs, or go to club meetings or sports practices.

Typically high school students like Brooke attend between four and seven 50-90 minute blocks of subject-specific classes. Theodore Sizer (1984), in his important book *Horace's Compromise: The Dilemma of the American High School* pointed to daily course schedules and the architecture, layout, and placement of classrooms as components that fragment the typical U.S. high school curriculum. National, state, and district level standards of learning guide teachers, parents, administrators, and students in what students should know and be able to do upon completion of each course. With the recent implementation and facilitation of high-stakes standardized tests in most states, teachers and students find their courses even more isolated and fragmented from other discipline areas. As one high school student reported, "It's like you go to entirely different worlds each time you change classes. Different teachers, different students, different rooms, and different information that in no way connect with your other classes or your world outside of school" (Sam as cited in Taylor, 1999, p. 71). Such curricular and physical fragmentation pose serious challenges for the design and implementation of new or alternative approaches to curriculum, instruction, and scheduling.

To combat this sense of fragmented and disparate curricula in schools, educational pundits call for interdisciplinary or integrated approaches to teaching and learning (Clarke & Agne, 1997; Sizer, 1984; Wolf & Balick, 1999). For the purposes of this book, *interdisciplinary* refers to the explicit recognition and connection of content and instruction from more than one subject or academic discipline in a teaching and/or learning experience. *Integrated* is defined as the merging of two or more course/discipline curricula. For example, one high school principal shared with us how historical events such as the French Revolution studied in world history class are separate or different somehow from the French Revolution studied in art or French class.[2] This way of thinking and teaching is a discipline-specific approach. Instead, in an interdisciplinary approach, teachers and students see, celebrate, and make explicit cognitive connections between and among their varied courses of study. An integrated approach is more holistic and involves students and teachers in combining discipline-specific ways of knowing to understand topics. **[We provide a more detailed discussion and description of interdisciplinary and integrated curriculum and instruction in Chapter 2.]**

The incorporation of interdisciplinary and integrated instructional approaches often appears futile to many high school art teachers who fear not only the loss of their already tenable positions but also hints of additional time and work to their already challenging schedules. Many high school art teachers witness the distortion of their art standards and curriculum by administrators who believe the arts should assist in the delivery of content in response to high-stakes tests. As experienced art educators who have taught in high schools at one time or another, we understand this situation all too well.

Students in many high school art classes range in grade and interest level and therefore rarely share classes or teachers in other disciplines. High school art classes are typically product and media-driven due to the demand for exhibition, competitions, art school and advanced placement portfolio construction, budget limitations, and advocacy issues. We rec-

Interdisciplinary refers to the explicit recognition and connection of content and instruction from more than one subject or academic discipline in a teaching and/or learning experience.

Integrated is defined as the merging of two or more course/discipline curricula.

Photograph by Donna Green, 2006.

ognize that these concerns are important issues if not obstacles to the implementation of interdisciplinary instruction in high school art classes. We share in the concerns of high school art teachers who already feel that their programs are viewed as subservient to other academic subjects. We support and value the learning experiences that are idiosyncratic to the study of art and believe strongly that, like other subjects in the school curriculum, the visual arts offer a body of knowledge worthy of sophisticated, in-depth investigation.

Art-full Teaching

High school art teachers teach art. Science teachers teach science. Math teachers teach math, and history teachers teach history. Right? Well, yes and no. As an example, let us look at a typical high school art lesson on perspective drawing.

In this lesson, the art teacher[3] shows students classic examples of the use of perspective in such works of art as Raphael's *School of Athens* and Leonardo daVinci's *Last Supper*. In addition to looking at the artists' use of such perspective techniques as linear perspective, overlapping, color intensity, placement, and size, the art teacher discusses the artists, the time in which they lived and worked, and the people their figures represent. In this learning process, the students deal with art history. If the art teacher happens to use a ruler to measure aspects of the work, a compass to determine the angles of the linear perspective in the work or in the students' work, they are, in essence, approaching math or at least using skills that are typically taught and reinforced in math. And if they go even further in their math connections, they might discuss the golden mean compositional technique and the Pythagorean theorem associated with Raphael's representation of Pythagoras.

Now you will notice that in this discussion, we did not say the art teacher was teaching history or math in this lesson. We said the teacher and students were dealing with, approaching, and making connections. In fact, art teachers do this all the time. They have to do this because "individual disciplines are not sufficient to approach significant issues and the expanding recognition of the advantages of project-centered work" (Beane as cited in Clarke & Agne, 1997, p. xii). Imagine that a math teacher came into the art class and used the implied linear perspective lines of Raphael's painting—discovered and marked by the students—to form a triangle that begins with Pythagoras, extends to the vanishing point

Figure 1.1. Raphael (1483-1520) *The School of Athens*, Stanza della Segnatura, Stanze di Rafaello, Vatican Palace, Vatican State. Permission of Scala/Art Resource, NY.

in between Raphael's central figures, Plato and Aristotle, and continues from the vanishing point to the point perpendicular with the starting point at Pythagoras—conveniently forming a general right triangle. (See Figure 1.1 and 1.2.). As the students measure the sides of this triangle, they create squares of each side to encompass more figures. In addition to working to understand the relationship of the size of the sides of the triangle, the students compare compositional elements in each of the squares. For example, the hypotenuse of the triangle forms the largest square with the largest number of figures as well as the largest area of architectural space. In this example, the art and math teachers may work together as a team to involve their students in observation and experience of formulating connections between the study of art and the study of math. Continuing this interdisciplinary process, students may work to compare the theoretical beliefs, teachings, and historical significance of the figures in the squares. In the process, they may also discuss how Raphael's painting expresses another important aspect of perspective—the way we see and perceive.

Figure 1.2. Illustration of an exercise involving the construction of a right triangle connecting the figure of Pythagoras with the vanishing point in Raphael's *School of Athens*. A right triangle is a triangle with an angle of 90 degrees. The sides of a, b, and c of such a triangle satisfy the Pythagorean theorem that states $a^2 + b^2 = c^2$. Photograph by Donna Green, 2006.

With this realization, the class might shift its focus to another painting, like Faith Ringgold's *Sunflower Quilting Bee at Arles*. In this quilt painting, Ringgold creates representations of a number of the greatest female figures in African American history, including Rosa Parks, Mary McCloud Bethune, Sojourner Truth, Ella Baker, Harriet Tubman, and Madame C. J. Walker. A study of Ringgold's work through the lens of perspective reveals an almost opposite approach to Raphael's *School of Athens*. Instead of using linear perspective, overlapping, size, color intensity, placement, or detail to create the illusion of depth, Ringgold places all of her figures on an equal plane and suggests various points of view. The viewer looks down on the quilt the women make while at the same time are at eye level with the figures. Further study of Ringgold and her African-American heritage suggests that she purposefully used these contradicting perspectives as contextual tools (Doyle, 2004). Ringgold chose to re-present the figures on a level plane with each woman as valid as those around her. This technique is filled with cultural implications and avenues for study and understanding. In addition, the study of these important African-American historical figures can be directly linked to high school social studies curriculum. In collaboration with social studies and history teachers, art teachers can initiate obvious interdisciplinary opportunities here, such as research projects, presentations, and links to social studies and history classes, as well as involvement with co- and extra-curricular clubs, school assemblies, and local agencies in the planning and execution of events or community projects related to the content of these works. In this example, students are involved in critical inquiry and research-based approaches to interdisciplinarity. **[Take a look at our discussion about project and problem-based approaches to interdisciplinary curriculum in Chapter 3 and Chapter 5 for more on inquiry and research.]**

Now, let us say that the students continue their study of perspective and use perspective techniques to create drawings of a commons area or public space in the school, such as the cafeteria, locker hallways, or library. For this assignment, the students create a representation of a specific space in which they include figures of the most influential thinkers, musicians, teachers, authors, sports figures, or other such people, to their own lives. They research images of these people and select information about them to share in their artists' statements. The art experience for these students becomes a project that requires research, writing, drawing, and observation. In this scenario the art students might also initiate another avenue of discovery by bringing to the class' attention the idea that the computer and the Internet suggest still another view of perspective—that of virtual reality. Students might bring in clips from movies like *The Matrix* or *Lawnmower Man*, popular music videos, or television commercials that feature images associated with how they envision the world of cyberspace. In the process, their findings could suggest another perspective artmaking activity that includes altered linear perspective complete with lines of computer code as in the film *The Matrix*, cross-contour grid drawings of their figures as in films like *Lawnmower Man*, and/or multiple points of view within one image as in such music videos featuring Körn and Eminem. **[See Chapter 9 for more on visual culture and art education and Chapter 8 for a discussion of computer and digital issues in art education.]** They may employ assistance in this activity from their physics and geometry teachers or texts. Keep in mind that in this scenario, the interdisciplinary perspective art experience is student-initiated; the suggestions we offer here are avenues of discovery that students might initiate on their own rather than prescriptive assignments from the teacher. **[See Chapter 5 for more discussion of student-initiated inquiry.]**

This interdisciplinary unit example is based on the concept of perspective in which students solve various problems through performance and collaboration. Upon completion of the unit, the teacher might include students in the assessment process through oral or written critique, performance, or exhibition/presentation of their work either within the school, at another site, or in an online computer environment. **[See Chapter 10 for more information regarding evaluation and assessment.]**

Art is an Essential Life Skill

On a daily basis, art educators face the interminable challenges of why the study of art is important and what professional opportunities are available in art. Historically many parents have discouraged their children from pursuing what they may stereotypically call a "starving artist" career because they assume such an avocation to be steeped in pain and poverty. Other than graphic design, some people believe that art careers are limited. Besides the many art-related professional opportunities including design, illustration, and education, the fact of the matter is that art, beyond being a livelihood, is an essential life skill. Like math, science, reading, and writing, the ways of thinking, looking, seeing, and making inherent to the visual arts are an integral part of life and living. Like pupils in math class, art students learn to solve problems. However, the exploration of contemporary and historical works of art as well as the use and investigation of media and techniques promotes inventive approaches to problem solving rather than obedient, blind adherence to prescriptive formulae. Art students learn to experiment much like scientists, but are encouraged by their teachers and assignments to take risks and stretch the limitations of forms, ideas, and media. Students learn and extend the language of art through critical interpretation of theirs and others works of art and the visual culture that surrounds them each day. In other words, art students learn to read and write visually, verbally, and textually.

The career opportunities available for art students, like those of math, science, and English students, may include such discipline-specific jobs as fine artist, designer, and teacher. But even more importantly, art is an essential life skill that permeates and enhances virtually each and every present or future occupation or profession. The interpretation and creation of visual artifacts—images, performances, installations, and other forms of visual culture—require the construction of symbolic representations, associations, relationships, analyses, reconstructions, explanations, arguments, and understandings of the world. Other modes of inquiry simply do not and cannot construct knowledge in such expansive ways as those integral to the study of the visual arts.

Observational skills are essential to diagnostic procedures in the medical field. Physicians and doctors must look for and recognize visual details in order to understand, ascertain, and pinpoint disease to formulate treatment. According to Yale University School of Medicine researchers, learning to analyze paintings helps medical students become better at diagnosing their patients (Sherman, 2001). The Yale Center for British Art project conducted by Jacqueline C. Dolev, Linda Krohner Friedlaender, and Irwin M. Braverman (2001) involved medical students in "Learning how to look at visual clues in a painting and then describe them visually to others," which "carries over directly to help sensitize [medical students] to interpret X-rays or the symptoms on the patient themselves" (para. 1).

The ways of thinking, looking, seeing, and making inherent to the visual arts are an integral part of life and living.

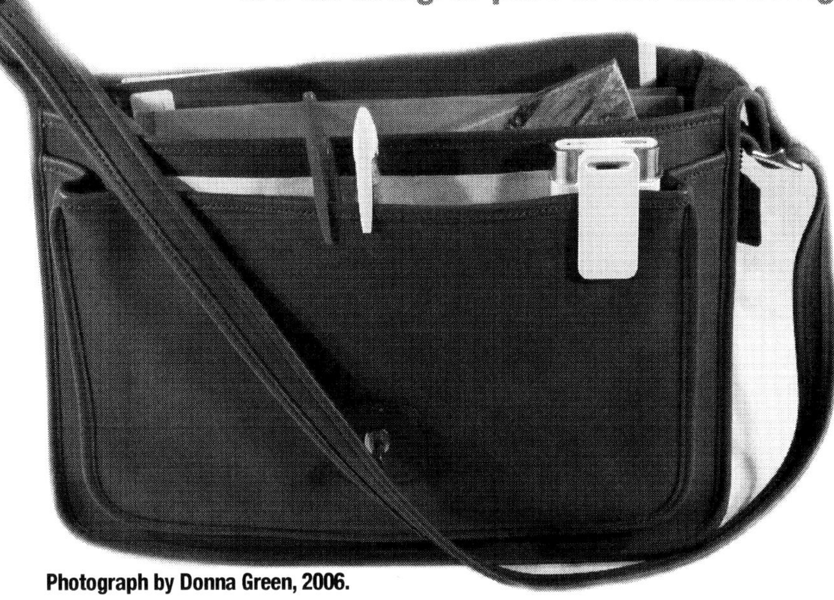

Photograph by Donna Green, 2006.

> Friedlaender says, 'It goes to higher critical thinking skills and enhances their ability to put visual clues together.' The pilot project then became a scientific study. 'We used a control group who had no museum experience and compared them to a group that did. And the group that had the museum experience scored a lot better on a follow-up test that looked at [medical] diagnostic skills,' Friedlaender says. The program works because it causes students to 'think inside the frame without any distractions,' Braverman says. 'The brain, in order to work, is organized to filter out unnecessary detail, and we need to overcome that in these medical situations. In every painting, every item, every detail has been placed there for a reason. The painting is the surrogate for the patient. A patient presents with signs and symptoms. A doctor is to find every one of those signs and symptoms and every one of them has to be explained.' (Sherman 2001, para. 12-15).

Police officers who solve crimes improve

NOTES:

Advocacy is an important part of what **we** do as art teachers. Know why as well as what you teach in your art class and make sure others know, too.

their observational and analytical skills through the examination of works of art (Byron, 2005). As reported in *The Wall Street Journal*, Amy Herman, head of education at the Frick Museum in New York City, led groups of newly promoted police officers in the examination of paintings to learn to broaden their assessment of police scenes before drawing or making conclusions (Byron, 2005). Herman required her police officer students to be more involved in detailed research in order to discover contextual clues. She also encouraged them to pay attention to detailed areas of paintings before making final interpretations or judgments. One of her police officer students reported that this experience directly affected his ability to solve a crime. In response, she reminded him to broaden his scope and include areas outside the immediate crime area. “In New York, the extraordinary is so ordinary to us, so in training we’re always looking to become even more aware as observers,” reported Diane Pizzuti, deputy chief and commanding officer of New York City’s police academy (Byron, 2005, p. A1).

NASA’s fine arts program recognizes the special ability of the artist’s eye to select and interpret what might go unseen by the literal camera lens (NASA, 1958). "The NASA Art Program uses the medium of fine art to document America's space program for 'the expansion of human knowledge of phenomena in the atmosphere and space. . . for the benefit of all humankind" (Woodfill, 2004, para. 3). Dating to the early 1960s, “NASA’s collection includes the work of artists ranging from virtual unknowns to the likes of Norman Rockwell and Andy Warhol” (McBride, 2004, p. A17).

"Mathematics seeks to describe reality by looking at the logical interrelationships between concepts. Through art, we experience reality in ways not directly accessible to reasoning, but which we find intuitively meaningful. . . .Both try to express fundamental "truths" about the nature of reality, seeking structure and symmetry within the complex universe in which we find ourselves" (McPhee, 1999, para 7). Many mathematicians are driven by a strong aesthetic sense of creativity. For example, researchers at the Hewlett Packard campus in Bristol, England bring together the worlds of advanced mathematics and fine art in their “Mathematics and Art Project” (McPhee, 1999).

The study of art promotes inventive and creative thinking that is essential to big and small business. A search through the library, bookstore, or World Wide Web for corporate employment issues lead us to dozens of books and articles on ways to promote inventive and creative thinking in the workforce. For example, Mark Fox (2003) referred to a survey of Fortune 500 CEOs that asked what they looked for when hiring employees and promoting managers. “A whopping 100 percent of the CEO’s mentioned creativity as one of the primary characteristics. In fact, almost 60 percent of the CEOs surveyed ranked creativity higher than intelligence” (p. 1, para. 8). Corporations enlist the help of such organizations as the ACA Group (A Consulting Alliance) to help them apply the creative process to business. This practice reveals how creativity and innovation are essential in today’s competitive business environment. The need to embrace change rather than simply manage change lies at the heart of successful organizations as workers learn to think in new ways, to develop creative responses to new demands, to be productively creative, and to stay ahead of the competition (ACA Group, 2004).

Art experiences heighten aesthetic value and awareness. In a July 2003 *WIRED* magazine article, “The Aesthetic Imperative: Why the creative shall inherit the economy,” Virginia Postrel wrote “Manufacturing and technology generate wealth only when they make matter and information serve human desire. Desire is the true source of economic value. Competition has pushed quality so high and prices so low that few manufacturers can survive on performance and price alone. To produce value, they must give customers something to please their sensory side" (para. 3). We recognize that the quest for beauty and taste have long been an accepted attribute or proposed result of art education. But, the fact remains that not all art is beautiful and any process that proposes to elevate taste and/or aesthetics is replete with problems. In fact, such an approach could hinder if not stifle art learning because the process could alienate important values of peoples, places, and ideas. With respect to aesthetic values and awareness, we prefer to recognize the importance of the study of art in heightening sensorial responsiveness and the elevation of appreciation for the visual world.

Elliot Eisner (2004) proposed that an aim of art education should be the preparation of thinking artists. The modes of thinking the arts evoke, develop, and refine, according to Eisner, relate to relationships that when acted upon: require judgment in the absence of rules; encourage students and teachers to be flexibly purposive; recognize the unity of form and content; require one to think within the affordances and constraints of the medium one elects to use; and emphasize the importance of aesthetic satisfactions as motives for work (p. 10). Such thinking is important for all students regardless of whether they would be labeled as gifted or talented in art making. Art "thinking" is key to preparing students to negotiate their visual world in meaningful ways.

The characteristics of flexible purposing may appear daunting, but seem to fit well with the inventive and creative ways in which many art teachers think, teach, and live.

Dennie Palmer Wolf and Dana Balick (1999) drew upon Clifford Geertz's (1973) argument for a "thick descriptive" way of teaching and learning that provokes teachers and students to approach and learn from cultures other than our own "through nuanced and empathetic accounts of life elsewhere, [rather than shallowly approaching a culture through] head counts, lists of holiday, or the names of rituals" (Wolf & Balick, 1999, p. 2). Wolf and Balick liken this attitude to a textured way of understanding that provokes a personal form of pleasure that encourages students and teachers to imagine, invent, and implement ways that make the study of the world thick with meaning. "Our knowledge is a montage of information gathered from our families, our lovers, our friends and colleagues, and our experiences. We need ways of working and learning that excite us" (p. 9). [See Chapter 6 for a detailed discussion of cultural issues related to interdisciplinary and integrated teaching and learning.]

Photograph by Donna Green, 2006.

Photograph by Donna Green, 2006.

Interdisciplinary models are simply raw materials from which high school art teachers may reinvent or reimagine their own meaningful and effective approaches to interdisciplinary instruction in and through the study of visual art.

Recognizing the value of *art-full* ways of thinking and knowing in a variety of situations and experiences can inspire meaningful interdisciplinary connections through various approaches and models. In other words, we believe that interdisciplinary ways of teaching and learning can be provoked through the thinking modes and skills inherent in art study and artmaking as well as through explicit connections between and among specific subject matter, ideas, and learning experiences.

Interdisciplinary Teaching and Learning Models

In the process of exploring and sharing varied interdisciplinary approaches to teaching in and through visual art, we discovered a number of basic models that we describe briefly in the next few pages of this chapter. Rarely, however, have we found an art teacher who conforms her or his teaching to any one of the following methods. After all, art teachers are highly inventive people, and are not typically comfortable with any one way of doing things. We applaud and celebrate this important and integral sensibility in high school art teachers and present the following models simply as raw materials from which high school art teachers may reinvent or reimagine their own meaningful and effective approaches to interdisciplinary instruction in and through the study of visual art. We will revisit these interdisciplinary approaches in the remaining chapters of this book.

Idea and Theme-Based. Units of instruction centered on ideas and themes that relate to human experiences offer teachers significant opportunities for interdisciplinary instruction. In order for meaning to be relevant to us, it is imperative that we connect it with other realms of experience both in and outside of school. In fact, "individual disciplines are not sufficient to approach [such] significant issues" as meaning (Beane as cited in Clarke & Agne, 1997, p. xii). Informed by art criticism theory and practice, "The key goal of interpretation is to construct meaning by building connections from a work of art to other texts" (Carpenter, 1999, p. 46). Because ideas and themes are rarely discipline-specific they can be incorporated easily into a variety of classes. Such interdisciplinary lessons might involve art students and teachers in art historical inquiry, artmaking, technique and media exploration. Based on ideas that relate to important human issues and experiences, thematic instruction in visual art demands the inclusion of content from a variety of other curriculum areas. Tom Anderson and Melody Milbrandt (2005) suggested that "The central instructional strategy for understanding ourselves and others through art and [the] visual culture in which it is imbedded is the use of themes" (p. 9). **[See Chapter 2 and Chapter 6 for more on idea and theme-based interdisciplinary instruction.]**

Problem-Based. Many art teachers consider problem-solving to be the essence of any and all artmaking experiences. Whether the problem is technique, media, or concept-based, artists and art students constantly work to find solutions through their creative process. Math, science, and English classes also use problem-based instruction in such activities as simple and complicated math problems, scientific inquiry and experiments, and creative writing assignments. **[For further discussions on problem-based curricula, see Chapters 2 and 3.]**

Project-based. It is important to note that we do not equate the term "project" with the works of art that students create in the art class. In this book, a project is a planned undertaking, task or problem in which students are engaged to supplement and apply their classroom studies. Just as in the case of discipline-specific learning, not all interdisciplinary experiences need to be project-based. However, as many art teachers will attest, hands-on work requires inventive thinking and the utilization of a variety of skills learned both in and outside the art class. Project-based interdisciplinary learning in and through the study of art can also include projects that may not lead to what is considered art. For example, art students might create a report, a garden, a fundraising initiative, or a performance. Projects might begin with a single discipline and gather knowledge and skills from other disciplines along the way. Projects in this approach to interdisciplinary instruction could also be service oriented. **[Take a look at Chapter 3, Chapter 4, and Chapter 7 for more on project-based learning.]**

Team teaching. Teamwork is essential to interdisciplinary and integrated learning in high school. Typically, high school teachers work independently and expertly in varied fields of knowledge. They, like art teachers, are passionate about their chosen field of study and are often practitioners of the disciplines they teach. For example, like art teachers who create art, many English teachers are writers. Many foreign language teachers are involved in cultural initiatives related to the language they teach through travel, cultural study projects, and ethnic organizations or festivals within their local communities. Some science teachers are actively engaged in scientific research in collaboration with local universities or research institutions. Math teachers may also work with universities and research institutions, or create new and exciting ways to approach the study of numbers. Therefore, just as the art teacher should be included and/or consulted when considering the study of art in other disciplines, the math and science teachers will have exciting, wonderful, and authentically meaningful ideas from their fields of study that art teachers can use with their students. **[See Chapter 2 and Chapter 4 for further details on team teaching.]**

Inquiry/research-based. Instead of depositing or covering information for students, inquiry/research-based approaches challenge teachers to use questioning tactics and analytic skills that will affect the direction of learning for their students. "In an inquiry classroom,

NOTES:

students become creators of knowledge rather than recipients" (Clarke & Agne 1997, p. 24). Like problem solving, an inquiry/research-based class may be built upon a particular problem or scenario from which students develop questions. They then formulate answers or responses to these scenarios through research. However, it is possible that in the process, student inquiry and research may change the direction of the class. We see such a development as an important, desirable, and challenging result of inquiry-based instruction. **[Chapter 3 and Chapter 5 offer more examples and discussions on inquiry/research-based instruction.]**

Collaborative learning. Students at various performance levels work together or "collaborate" in small groups toward a common goal. Collaborative learning provides opportunities for students to learn with and from each other. Proponents of collaborative learning claim that the active exchange of ideas within small groups increases interest among the participants and promotes critical thinking (Johnson & Johnson, 1986). Like teams of teachers from varied disciplines, teams of art students also bring alternative ideas and connection possibilities to their study and to their classmates. That is, because students in the typical high school art class represent different ages, grade levels, interests, and career goals, collaborative learning is inherently interdisciplinary. **[We explore collaboration with respect to interdisciplinary curriculum and instruction in Chapters 4 and 5.]**

Photograph by Donna Green, 2006.

Student-initiated learning. Teenage students like to control things. They like to have a voice in connection with what they do and think about in and outside of school. So, when teachers encourage students to provoke and initiate learning opportunities and interdisciplinary possibilities in their classes they make feasible exciting and effective learning situations. That is, when teachers actively and genuinely listen to students without thinking of their own agenda in the process, they provide exciting interdisciplinary possibilities. In fact, many of the most effective interdisciplinary connections that we have experienced were the result of student ideas. Simply asking students to explain how the content they have studied in the art class reminds them of something else, can take a discussion in directions that exceed the expectations of a planned unit of instruction. Of course, the teacher must always be available to guide and make sure student-initiated inquiry provokes valid learning opportunities. **[Take a look at Chapter 5 for more on student initiated learning.]**

Visual Culture-based. Every day high school students are exposed to numerous visual images, advertisements, and media. For most teenagers, logging onto the World Wide Web, playing video games, sharing pictures through their cell phones, and watching television are basic aspects of daily life. As high school students engage with these and other forms of material and visual culture, they draw upon knowledge about a variety of ideas, information, and experiences. "Challenging the old constructs

Teenage students like to control things. They like to have a voice in connection with what they do and think about in and outside of school. So, when teachers encourage students to provoke and initiate learning opportunities and interdisciplinary possibilities in their classes they make feasible exciting and effective learning situations.

of knowledge about the visual arts leads to at least some degree of interdisciplinarity. The 'real' of the visual arts overlaps with other school subjects. Art education should help students know the visual arts in their complexity, their relationships as well as their independences, their conflicting ideas as well as their accepted objects, and their connections to social thought as well as their connections to other professional practices" (Freedman, 2003, p. 19). **[In Chapter 6, Chapter 8, and Chapter 9 we discuss further the roles of visual culture within the lives of students and the high school art curriculum.]**

Experiential service-learning. Learning through experience is at the heart of what students and teachers do in the high school art classroom. When that learning moves outside the classroom and involves others in meaningful ways to access new knowledge, interaction, and reflection, it is experiential. That is, experiential learning is anchored in and/or brings real-world meaning to in-school learning. Experiential service-learning provides a way for students to engage in activities that link together and apply knowledge they gain from the study of various disciplinary subjects while they help, assist, reflect, and make a positive difference in their world. **[In Chapter 7 we discuss service-learning in detail.]**

Computer-centered. Within the past 20 years, the impact of computer technologies on the field of education has been huge. Information is now literally at or under our fingertips and in front of our noses through online interactive resources, local access networks (LAN), and computer disks. With software programs teachers and students create, monitor, grade, communicate, exhibit, plan, collaborate, play, and shop, as well as find their way from one point to another, submit homework, and take attendance. "In the process of reconfiguring the visual arts, advanced technologies have changed what it means to be educated" (Freedman, 2003, p. 22). Computer technologies challenge and provoke new and exciting interdisciplinary possibilities every day. They are tools to think with, make with, communicate with, and link with in a variety of settings and forms. **[In Chapter 8 we talk more about computer and digitally-centered learning.]**

Flexible assessment-based. Clarifying the goals for what students should know and be able to do upon the completion of a unit of instruction, a course of study, or a grade level is essential to authentic and successful interdisciplinary teaching and learning. One of the goals of assessment is to organize learning so that it is visible and can be documented. In this assessment-based process however, John Dewey's (1938) "flexible purposing" or the value of recognizing new and emergent possibilities for learning and knowing is intrinsic to art education. As Elliott Eisner (2004) cautioned "The kind of thinking that flexible purposing requires thrives best in an environment in which the rigid adherence to a plan is not a necessity. As experienced teachers well know, the surest road to hell in a classroom is to stick to the lesson plan no matter what" (p. 6). The characteristics of flexible purposing may appear daunting, but seem to fit well with the inventive and creative ways in which many art teachers think, teach, and live. Although developing flexible yet assessable goals for interdisciplinary instruction may appear to be interminable, authentic attempts by teachers to do so are crucial. A flexible purpose is a purpose just the same and every lesson of every unit must have a purpose and goal. Indeed, asking oneself, "Is this worth knowing and why?" should be an essential qualifier for everything we do in education. A flexible assessment-based approach to interdisciplinary high school art teaching demands that teachers and students from multiple disciplines formulate, reformulate, and reflect on the degree of attainment or effect of important goals and aims for each learning experience. **[See Chapter 10 for more on evaluation and assessment.]**

Suggested Discussion Questions and Activities

1. Earlier in this chapter in the section "Art is an Essential Life Skill," we refer to a few important life skills that may be enhanced or learned during a student's study of art. Read Elliot Eisner's (1998) article in *Art Education* entitled "Does Experience in the Arts Boost Academic Achievement." Compare Eisner's suggested art study outcomes with what we have discussed.

2. How realistic is it for students to learn inventiveness or creativity? What kinds of activities would you organize that would promote inventive skills? How would you challenge your students to be critical thinkers? (See Lampert, 2006) ask questions & set-up problems for them to solve

3. Referring to the example unit of instruction on perspective, what other interdisciplinary possibilities can you see in this unit? How can teachers of other disciplines expand this unit? What works of art would you use in place of the ones we have described?

4. In this chapter we have offered a number of approaches to interdisciplinary instruction. What if a student discovers a connection between what she or he is studying in another class and what is being studied in art? What approach(es) would you take to emphasize that connection for the students?

5. What would a science or math teacher contribute to one of your favorite art units or lesson plans? What would you contribute to a science or math unit? Where would you begin to develop a unit of instruction with a math or science teacher? Why?

Notes

[1] Brooke is a fictional blend of students in author Pamela G. Taylor's memory of teaching at Christiansburg VA High School 1988-1997.

[2] Mr. Miller, former principal of Auburn High School in Riner, VA shared this concern with author Pamela G, Taylor during an arts curriculum restructuring meeting in November 1996 in Christiansburg, VA.

[3] This unit was taught by author Pamela G. Taylor in 1997 at Christiansburg VA High School.

Chapter 2

WHAT ARE WE TEACHING AND WHY?

Jeremy[1] was in high school and struggling in algebra class. Frequently, he asked, "Why do I need to learn/know this?" His favorite class was art and design. One day, as he flew his hand-built remote controlled airplane in the local park, his algebra teacher strolled by. In the conversation that developed, the teacher stated his surprise over Jeremy's choice of hobbies. His teacher, Mr. Thomas, knew that building remote-controlled airplanes requires a great deal of measuring and calculating. Mr. Thomas wondered how a student like Jeremy, who had such a hard time grasping math concepts and seemed to spend all of his time drawing, could engage in and be successful at such a mathematically demanding hobby. The algebra teacher did not connect in his own mind much less in his teaching that such students' interests and hobbies outside of math class were related to the work they did in class. As a consequence, Jeremy assumed that his algebra teacher believed he was not smart enough to understand math concepts much less build remote-controlled airplanes. What could have been a teachable moment and/or an epiphany for both Jeremy and his algebra teacher instead was lost or dismissed. As a result, Jeremy became further disengaged in the class. And like many high school students, Jeremy decided that he could just "get by" if he simply took an easier math class. In the process, Jeremy missed an opportunity to understand the larger contexts of learning.

Interdisciplinary/Integrated — Mapping & Standards

Big Ideas — Key Concepts & Skills — Essential Questions

We know generally what is to be learned through and in algebra. We know theoretically, what should be gained and the connections from algebra to other content. Although many of us may feel that we manage to go through life perfectly fine without applying algebra skills to real world situations, in the long run we know that students in this situation short-change their education. Students need to establish context, content, and simple relationships between what they find in their textbook pages and course content with what they experience and know about real life. They need to ask questions freely. They deserve to understand why they need to know "this," why their teacher is teaching "that," and what is really being taught and learned in their education experiences.

We know that visual arts help students discover more about their identity, culture, and community. We also know that visual arts employ diverse ways for people to know their world physically, emotionally, cognitively, and spiritually. Among our goals as educators is to prepare students for their present and future social lives, work, and citizenship. If we are to relate visual arts to life, we must do so in ways that anticipate the questions: Why do I need to know this? Why are you teaching this? What is really being taught and learned?

To help students prepare for their present and future lives, art educators must relate the school curriculum through inquiry-based learning that focuses on life-centered issues relevant to the lives of students. This approach must also relate to the learning in a general course of study that integrates and connects disciplines, and makes curriculum relevant and meaningful. Such an approach will answer the questions we have raised, and more.

Relating to Terms

Although the terms interdisciplinary curriculum and integrated curriculum are often used interchangeably, they are not the same. Interdisciplinary curriculum requires subject areas to remain distinct and curriculum planning to be based on large concepts, themes, issues, problems, or questions that are meaningful to the lives of the students and that are shared, or that crossover subject or discipline areas. In the development of an interdisciplinary curriculum, these overarching and meaningful themes, concepts, issues, problems, and questions are taught and dealt with usually in contained classrooms by various subject teachers who are involved from the perspective of the subject or discipline that they know. Generally, teachers in middle, high school, and higher education use the interdisciplinary curriculum format because the traditional structure of the school is more conducive to this form.

In integrated curriculum, the overarching and meaningful themes, concepts, issues, or problems are also planned for and taught, however, the subjects or disciplines are no longer separate. Instead, the subjects or disciplines are woven together to make a more gestalt whole. One often cannot tell, for example, where geography and art begin and end. An integrated curriculum takes even more collaborative planning and time than an interdisciplinary one and, for this reason integrated curricula are enacted more often at the elementary level where the teachers have a shared culture already in place. Typically, integrated curricula are team-taught in 2- to 3-hour-long classes.

Interdisciplinary and integrated approaches are valuable and they move toward educational reformation that supports collaboration. When teachers at any level base their teaching on inquiry and facilitate problem solving rather than mere fact acquisition, they help students understand, connect, and interpret overarching and meaningful knowledge of themes, concepts, issues, problems, or questions in a rich and complex manner.

Because interdisciplinary education generally falls outside the norms of "school or school system culture," support is necessary from key individuals and groups in order to launch and sustain such programs. Advocacy and the support of department heads, principals, curriculum directors, superintendents, and/or school committees are essential to interdisciplinary and integrated teaching. We understand that tradition-bound colleagues may feel threatened by what they consider the unknown, a risk to their discipline-specific way of thinking, or a loss of control or power. We are all defensive of our disciplines. An inhospitable community may exert its own forms of pressure. Parents in particular may discourage their children from signing up for interdisciplinary electives or may complain about non-elective programs that seem confusing or eccentric (Jacobs, 1989). Teachers need to emphasize to both internal and external communities how integrated and interdisciplinary curricula prepare students to be connected, critical thinkers, enable them to relate concepts and skills to other areas of life outside school and prepare them for adult life-long learning. Such an approach to critical thinking also requires students to analyze and evaluate as they "sort out their own reactions and articulate them through the medium at hand" (Fowler, 1996, p. 48).

Relating in Different Ways

Since 1991, general education theorists, Heidi Hayes Jacobs and Rebecca Crawford Burns have advocated that curricula be connected through big ideas, key concepts, and essential questions. To take their notions further, we add that the **process** of developing lessons and units of instruction through big ideas, key concepts, and essential questions is the most important step. Through collaboration with other teachers, rather than in isolation, these processes promote and nurture discussion for teacher understandings of cross-discipline planning. Inherent in this process must be **reflection** that fleshes out multiple perspectives, ambiguities, nuances, conflicts, complexities, skills and interests of teachers and students.

One way of making connections with other teachers is through the school calendar. Jacobs (1997) stated that all teachers, no matter what discipline have the school calendar in common. Although most teachers may not know what other teachers are teaching, they do have a time schedule to which they must adhere. Collaboration could begin as early as during the opening weeks of school as teachers plan unit themes, content, skills, and projects for the year. Jacobs referred to this approach as mapping. According to Jacobs (1997), the purpose of mapping is to reveal what is actually being taught. Mapping is a process of outlining major components of curricula: objectives, skills, content, projects, and performances that are designed and assessed. (See Figure 2.2 and 2.4) In this case, teachers include themes in their yearly plan that may be a part of the content or

If we are to relate visual arts to life, we must do so in ways that anticipate the questions: Why do I need to know this? Why are you teaching this? What is really being taught and learned? Is this worth knowing?

art as a tool rather than a discipline
about the process rather than the outcome. Skills for life. Not always about the project, but about the collaboration.

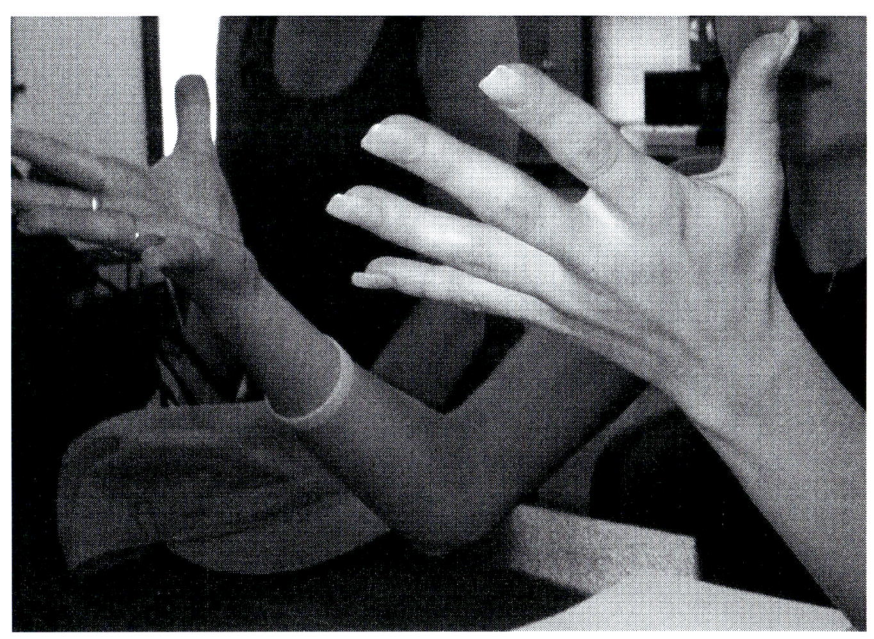

projects and performances. Commonalities and repetitions reveal themselves in such cross-disciplinary calendar mapping. [For more on mapping and curriculum planning, see Chapter 8.]

Coordinating skills and content with the school calendar. Let us look closely at the example of Ms. Sharn, a high school art teacher, and two of her colleagues: Mr. Greg, life science teacher and Ms. Lin, home ecology teacher.[2] Ms. Sharn began each new school year with a unit on gardens specifically referring to Claude Monet's garden at Giverny and the J. Paul Getty Museum's garden designed by Robert Irwin. Her students explored design elements, functions, sculptures, and environmental works. Mr. Greg involved his students in studying and creating a garden at the beginning of the school year also. His science students charted and made scientific observations of the growth progress of their garden in their journals. During the spring semester, Ms. Lin emphasized the historical aspects of gardens in her human ecology classes.

Ms. Sharn, Mr. Greg, and Ms. Lin realized that together they could create an interdisciplinary unit with projects that built upon each other rather than ones that existed in independent, isolated, or separate parts. Their mapping process was an example of the complex relationship among calendar year, skills, and content. For example, Ms. Sharn and Mr. Greg typically taught their garden units at the beginning of the school year and Ms. Lin engaged her students in this topic in the spring. Upon closer scrutiny, the three teachers realized that the skill sets taught in their respective garden units were not dependent on a particular time during the school year. In other words, the garden unit did not require building upon specific skill sets or concepts that were taught prior to these units. Therefore, the three teachers could change their plans so that they taught their garden units at the same time of the calendar year.

Ms. Sharn's students explored the Los Angeles J. Paul Getty Museum's Getty Center garden created by artist Robert Irwin as a contemporary example of how foliage and materials accentuate the interaction of people with light, color, and reflection. (See Figure 2.1) Irwin's philosophy was that the gardens were in a continual state of flux. Irwin's statement, "Always changing, never twice the same," was carved into the plaza floor, and reminded visitors of the ever-changing nature of this living work of art. This belief influenced the Getty garden designers and viewers as the gardens were constantly transformed both naturally through growth and physically through the gardeners' planting and clearing. The content of Ms. Sharn's unit focused on these relationships between human beings and the natural environment in which they live and included their form, function, interplay, and impact on each other. In Mr. Greg's science class, the students looked more scientifically at these relationships through such concepts as climate, photosynthesis, and botany. For example, the walkway in the Irwin garden descended to a plaza shaded by arbors covered in bougainvillea. Said to provide scale and a sense of intimacy, students discovered that bougainvillea was a vigorous, evergreen with woody vines and spines. The vibrant color of this vine came not from the small white tubular flowers, but from the 3 large paper-like bracts that surrounded each flower. Students learned that bracts were small, sometimes scale-like leaves, usually associated with flower clusters. And in Ms. Lin's human ecology class, students discovered that Admiral Louis de Bougainvillea discovered the vine in Brazil in 1768 during his long journey to the Pacific Ocean.

The unit goal for the art class was the demonstration of an understanding of design methods and skills including the use of color and light in the form of an actual garden creation. Ms. Sharn's students created a miniature garden design replica appropriate for the community's geographic location.

Mr. Greg's science students were involved in critically examining the garden design as well as documenting and analyzing its growth. The goal for Mr. Greg's science class was a demonstration of the use of appropriate tools and techniques for scientific inquiry including gathering data, and thinking critically and logically about relationships between geographical climate, soil, and botanical life cycles.

Ms. Lin's human ecology students traced the origins of many of the plants and flowers used in the garden design as well as compared the purposes and historical influences of similar garden designs. The goal of the human ecology garden unit was to demonstrate an understanding of the contemporary and historical relationships between human beings and the natural environment through a detailed analysis of design, plant choices, geographical location and human interaction.

Ultimately, Ms. Sharn's students created the first design for the garden. Mr. Greg's class revisited their understandings and explorations of gardens according to scientific information and redesigned the plans for both classes. Ms. Lin's students investigated the historical significance and influence of garden planning on human interaction. And finally, all stu-

Figure 2.1. Robert Irwin, *Central Garden*. J. Paul Getty Museum, Los Angeles. Photograph by Alex Vertikoff. Permission © J. Paul Getty Trust.

dents were involved in the creation, study, and maintenance of an actual garden.

In this approach, Ms. Sharn, Mr. Greg, and Ms. Lin sequenced the topics and units in their individual curricula to relate with one another and the similar concepts they shared were taught in concert with one another while each of the subjects remained separate (Fogarty, 1991). Once the team of Ms. Sharn, Mr. Greg, and Ms. Lin wrote the skills, content, project, performances, and objectives for their interdisciplinary unit, they reviewed each other's responses. They looked for repetitions, overlap, and skill building.

Review and repetition. Ms. Sharn's art students revisited basic principles of design as well as color theory when studying the garden designs of Monet and Robert Irwin. One focus in developing interdisciplinary or integrated curricula should be the possibility for relevant redevelopment. We know that there is instructional value in designing curriculum so that students have opportunities to repeat concepts. Revisiting a novel in English class can expand on concepts or revisit them when the students read at a different maturity level. Similarly, an advanced or relevant review of basic color theory and relationships in high school art provides students with an opportunity to use their knowledge of color in their creation and analysis of works of art. Crucial however, is the need to take a more advanced approach to the concept being repeated. Re-reading a novel is not the same as creating yet another color wheel. For example, just as 9th grade students who read a novel for the second time would not explore the same terms they learned in 4th grade, reviewing color in a high school art class should not require the same approach as in elementary school. When reviewing color at the high school level, we advise art teachers to approach this through the creation of quick color value and wheel exercises that can be referred to during the exploration and creation of meaningful works of art rather than spending valuable time on ornate color wheel creations. Ms. Sharn's art students reviewed color theory by identifying specific areas in photographs of Monet's and Irwin's garden where relationships between adjacent colors of flowers

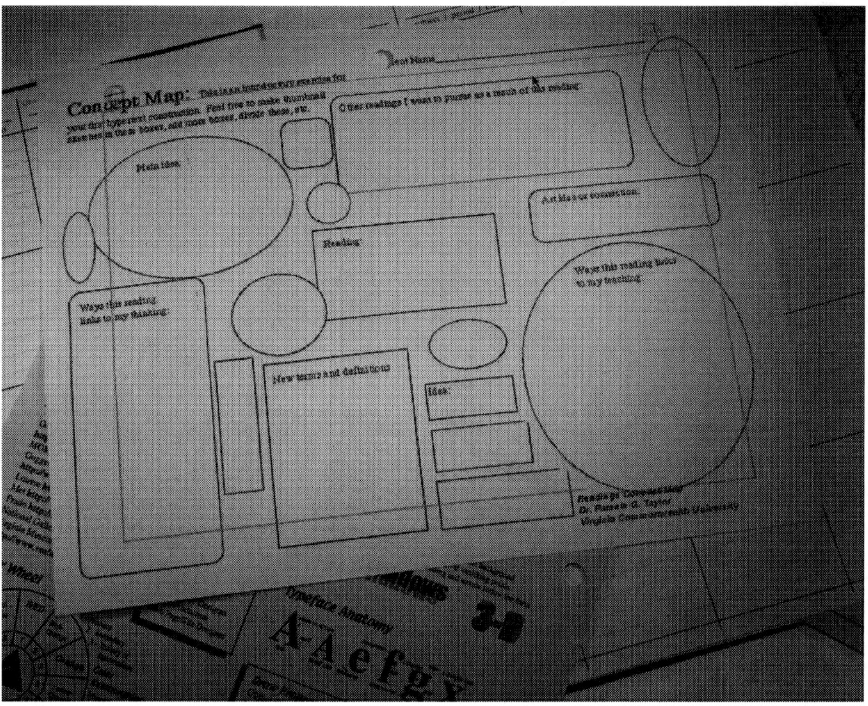

Figure 2.2. Mapping is a process of outlining major components of curricula: objectives, skills, content, projects, and performances that are designed and assessed Photograph by Donna Green, 2006.

functioned for example, to balance, focus, create directive devices, and/or convey mood or emotion. Using the computer, they created digital snapshots of these areas and linked them with images of color wheels and value scales in their digital journals. And they explained for example, how the use of analogous warm colors along the pathway directed viewers through the garden toward the ethereal arbor of bougainvillea. In this way, Ms. Sharn's art students applied the color theories they learned prior to this class. Through the redevelopment of what could be called their color skills, they purposefully chose specific flowering plants according to color relationships and design purposes for their own garden.

Overlapping concepts and skills. Depending on the communicative styles

Figure. 2.3. Students in Sara Gant's classes created an interdisciplinary star book containing poetry, letter-writing, descriptive writing and history. Permission of the artist.

of teachers, both sequential and shared interdisciplinary approaches support the development of curricula with another person. An example was shared with us by Ms. Sara Gant, art teacher at Northside High School in Jacksonville, NC. [Teacher contribution.] Her unit illustrated an interdisciplinary approach when concepts and skills of two or more disciplines, in this case art and English, were combined. (See Figure 2.3). Ms. Gant explained:

> In the fall semester of 2001, students came to art class with a project due for their English class. It was an autobiographical project, and consisted of scrapbook-type works that the students created by hand. The students were proud of these books, and other students asked to look at them with interest and eager anticipation. The pieces contained decorative artwork, photographs, and different sections about the students' lives, likes, and dislikes. This was obviously a very successful teaching tool. English teacher Ms. Jennifer Whaley discovered that personal and creative work was meaningful to her students. Ms. Whaley created this assignment and taught it for many years. I, the art teacher took the concept and assigned a similar project to Art I students—create a book about an artist that hopefully would be a valid teaching tool to share with the rest of the class (email communication, February 5, 2005).

Based on a shared assignment and with similar skills, the art and English teachers combined their assignments and some of their instruction to make the experience connected and meaningful, but not duplicated. This type of thinking on the part of the teachers was necessary in order to move from a "my classroom" mentality to an "our students" way of thinking and teaching. Similarly, in Ms. Anna Ursyn's unit entitled "Journey to the Center of the Earth," computer graphics activities are integrated with physical geology-related concepts, events and processes, language arts, and connotations pertaining to cellular biology. [Teacher contribution.]

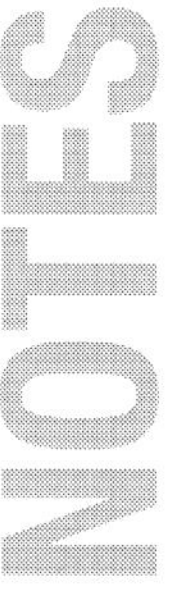

Essential questions organize the structure of the myriad elements that are a part of curriculum—connecting the activities.

> Jules Verne's novel serves as a starting point for an art project about physical makeup of the Earth and forces affecting its structure. The challenge is to visualize this scientific concept and create a personal artistic solution as a response to the description of the Earth structure, first in the scientific then in the literary terms. (personal communication October 4, 2004).

Ann Gerold, art teacher at Mounds View Area learning Center in Shoreview, Minnesota shared with us a "body art" unit in which her art students drew upon their biology class studies to create costumes representing specific bodily organs. Her students presented and performed for area elementary school students who were studying human anatomy. [Teacher contribution.]

Ms. Sharn's garden unit, Ms. Gant's book unit, Ms. Ursyn's earth unit, and Ms. Gerold's body-art unit are examples of two differing approaches to interdisciplinary connections. In the garden and body-art units, art, life science, biology, and human ecology shared a common topic during a certain time of the calendar year. In the book and earth units, art and English or art and science worked together to enhance a shared assignment. Another third approach involves the integration of curricula through disciplinary commonalities related to concepts.

Relating Conceptually

Developing curricula through essential questions may more clearly illustrate why something is being taught. That is, a curriculum designed from essential questions answers "Why do I have to learn this?" To combat the problem of integrated curriculum being just a series of activities, much care should be taken to determine why certain activities are being pursued, what is being taught, and how it is related to the content of the disciplines and to necessary skills. Essential questions organize the structure of the myriad elements that are a part of curriculum—connecting the activities. For an essential question to be successful it should be: easy to understand, broad, reflective of conceptual objectives, sequential, and visually posted in the classroom.

According to Tom Anderson and Melody Milbrandt (2005), "Researchers have found that students learn better and more deeply when they take up powerful ideas, with units organized around key supporting concepts, than when they learn mere facts or techniques" (p. 7). Key concepts are the centers around which meaningful curricula are developed. At a general level, key concepts represent significant aspects of life such as identity or social codes, and they reflect the complexities, ambiguities, and contradictions of our culture. Complexity, ambiguity, contradiction, paradox, and multiple perspectives also characterize good key concepts. But, key concepts can also combine concrete and abstract elements in a provocative way. Key concepts do not suggest ready answers so much as they invite inquiry. They help students ask questions about the world and about themselves and they call for an understanding of more than one point of view. The development of key concepts and essential questions is critical to meaningful integrated/interdisciplinary instruction.

For an example, let us revisit and expand the garden concept in the unit designed by Ms. Sharn, Mr. Greg, and Ms. Lin. If we explore the garden concept through the big idea of environment, the content in art and other disciplines may be investigated. Because it is a large topic, we must focus on issues related to environment, or what Jacobs (1997) referred to as key concepts. Ms. Sharn's garden theme explored the way people are influenced by their environment, but this concept was never overtly stated nor explicitly taught. Through group discussion and negotiation processes

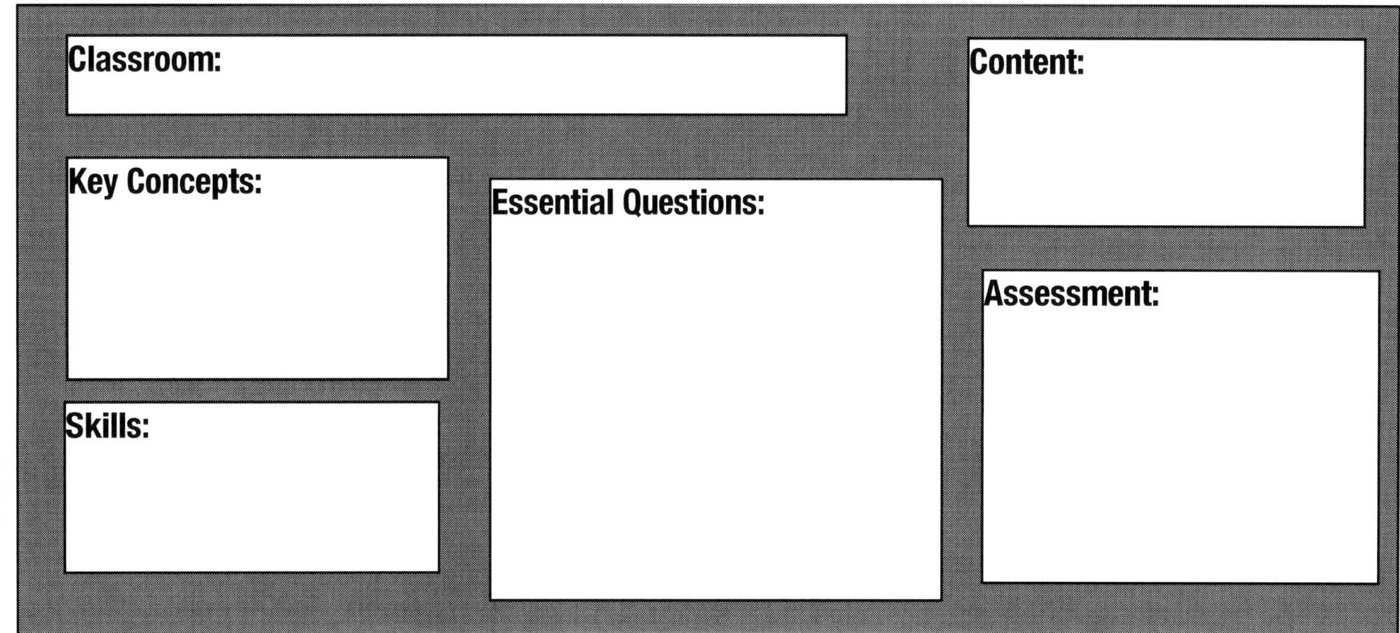

Figure 2.4. Planning with Key Concepts and Essential Questions

that take student thinking in multiple perspectives, skills, interests, and key concepts will emerge. The garden theme expands into a key concept this way: the environment affects all individuals and we all affect the environment. With just this change, the health teacher could also link to the key concept and address conditions such as stress, environmental illnesses, and healthy lifestyles. The math teacher could connect the use of diagrams to convey statistics about the environment and health data and the use of mathematical formulas to determine environmental living spaces.

The characteristics of good key concepts are foundationally necessary in order to make genuine connections and to stimulate inquiry-based learning. A good key concept must possess significant aspects of life, which is determined through classroom discussions. Good key concepts potentially provide opportunities to explore the complexities of cultures and multiple points of view. Further, good key concepts combine abstract and concrete elements and may also capture students' interests and needs.

Once key concepts are determined, teachers should turn those statements into essential questions (Jacobs, 1989). Using environment as our example, the essential questions might be:

What is an environment?
How has the environment affected people throughout time?
How have people affected their environment throughout time?

Photograph by Donna Green, 2006.

By turning the key concept statements into essential questions, teachers establish a conceptual direction for inquiry-based exploration. Essential questions serve to reveal the deeper connections of the key concepts. These questions also provide students and teachers with opportunities to investigate what we do not know as revealed through brainstorming exercises or further discussion. A key concept can be turned into essential questions by re-phrasing the concept into a series of questions. Essential questions give conceptual direction to student thought and invite investigation, research, and analysis. As McTighe and Wiggins (2004) stated, essential questions have no definitive answer, are designed to provoke inquiry, address conceptual foundations of disciplines, and stimulate rethinking of concepts, assumptions, and prior knowledge. That is, essential questions inspire divergent rather than convergent thinking.

When teachers center units of instruction on concepts and themes that relate to human experiences they offer their students significant opportunities for interdisciplinary instruction. Because concepts and themes are rarely discipline specific we have found it natural to develop interdisciplinary units of instruction using concepts, essential questions, and themes. All of these approaches can be incorporated easily into a variety of classes. For example, lessons may involve art students and teachers in art historical inquiry, art-making, technique, and media exploration. Based on concepts that relate to important human issues and experiences, thematic instruction in visual art demands the inclusion of content from a variety of other curriculum subjects.

Tips to Remember

1. Content must stimulate answering the question—the essential question is the foundation of the curriculum and the inquiry.
2. An essential question determines activities—without this the curriculum could be just a string of activities.
3. Projects, performances, and assessment must also relate to the essential questions.
4. Use no more than three essential questions per unit.
5. Questions should be easily understood by the students and dis-

played in the classroom and hallways.
6. All of the skills and content from each of the disciplines represented in the integrated curriculum should be displayed in all classrooms involved. Doing so helps illustrate collaboration and connectiveness.

Relating to the Standards

One important consideration for teachers is the alignment of the curricula they design with school and state curriculum guidelines. State standards may serve as a framework that structures the existing curriculum or standards may be elements that are already embedded within the units. In either case, the key concepts and essential questions should be relevant to the students' lives, as well as meet the needs of national, state, and local standards and assessment requirements. **[For more on standards, assessment, and evaluation, see Chapter 10].**

By creating an artist book, each student in Ms. Gant's art book unit consciously chose a theme and a style of their own that they adhered to throughout their book. This visual style was portrayed through their particular use of the media, techniques and processes that they employed. Therefore, Ms. Sara Gant's unit on art books addressed the State and National Frameworks for Secondary School Instruction Content Standard #1: Understanding and applying media, techniques and processes which stated, "Students apply media, techniques, and processes with sufficient skill, confidence and sensitivity that their intentions are carried out in their artwork. Students conceive and create works of visual art that demonstrate an understanding of how the communication of their concepts relates to the media, techniques and processes they use" (National Standards for Arts Education).

By creating visual pages of art and text containing at least eight descriptive stylistic traits and character traits, as well as creating an original poem in response to the readings about her or his artist, Ms. Gant's students met the North Carolina English Language Arts Curriculum Competency Goal #1 that "The learner will express reflections and reactions to print and non-print text and personal experiences." The goal required that students were able to: "Respond reflectively...to a variety of expressive texts...in a way that offers an audience an understanding of the student's personal reaction to the text" (North Carolina Standard Course of Study).

It is clear then, from looking at the ways that this project met goals for several disciplines, that students benefited from the thought, preparation, and interdisciplinary connections that were made during the creation of their books.

A process that helps simplify and connect curricula to standards is called scan-and-cluster (Drake & Burns, 2004). Scanning is just that: looking through the same subject areas and then clustering them according to the similarities. Teachers must remember to be aware of the standards that address skills and the ones that address content.

Drake and Burns (2004) suggested that teachers first select standards at the appropriate grade level for the areas that will be integrated. Next, teachers should identify one skill that is interdisciplinary and involves complex performance such as research or communication. Teachers who are interested in interdisciplinary connections should then discern subset skills. For example, research subset skills would require students to locate resources, organize, and cite references. Then, teachers should

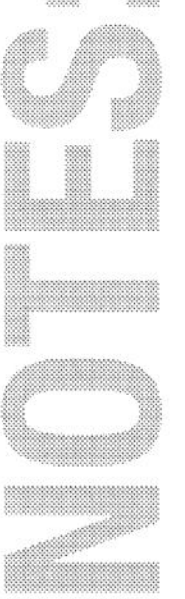

scan the standards in one subject area to locate specific standards and then cluster the standards into similar chunks. To complete the process, teachers would repeat this procedure for each subject area. As additional subjects are examined in this way, similarities across the disciplines will emerge. Whether skills or concepts are the focus, this process helps commonalities emerge across the disciplines. This process also helps teachers from multiple disciplines begin to understand standards and help solidify connections and accountability across the larger school curriculum. Drake and Burns (2004) stated that the following areas were most often evident in curricula: research, scientific method-inquiry, problem solving, communication, design and construction, presentation, comparison, prediction, and making charts and graphs. In the case of the garden/environment, all three teachers used not only design and construction but also scientific method-inquiry.

Relating

A general desire for most teachers in any discipline is the making of meaning. We know that if we make connections, present material in multiple ways for student consumption, employ an inquiry-based approach, and meet standards while making them "come alive," student learning will be enhanced. Drake (1998) stated that if students learn how to transfer knowledge from one area to another they are more likely to become lifelong learners. The arts can be the leaders in this area because the arts do not have to be taught in isolation to be taught well. This characteristic makes the arts conducive for integration. Art educators have an opportunity to be a leader for integrated curricula in their schools. This perspective does not mean that there will not be challenges and an initial investment of time on the part of teachers. Nevertheless, we believe that with such an approach, the outcomes and possibilities for student success and learning are endless.

Suggested Discussion Questions and Activities

1. Your school has just discovered a newspaper article reporting that the scores on the state standardized testing in math are again below state standards and your school is now being watched carefully. The principal has decided that measurement is the school-wide theme. As the art teacher how do you proceed and connect to the other classes?

2. Taking into consideration the concept of curriculum as a process that incorporates reflection and mapping, you are in charge of teaching all of the art courses at a small rural high school. The teachers view your role as supportive; that is they see you as the one who organizes the poster contests and keeps the alternative students quiet. The senior year team has decided their focus is on scientific research. How will you connect? Develop a set of key concepts and essential questions.

3. It is winter and you usually concentrate on preparing your senior students to develop their portfolios for the annual university art school portfolio review. Develop key concepts and essential questions that would go beyond simply creating a portfolio. Review the ideas and research in Mary Stockrocki's (2005) *Interdisciplinary Art Education: Building Bridges to Connect Disciplines and Cultures.*

Notes

[1]Jeremy is a fictional blend of students in author Christine Ballengee-Morris' former secondary school classes.

[2]This story is a fictional representation and extension of the work done by three teachers involved in an interdisciplinary research project with author Christine Ballengee-Morris in Ohio 1997-2002.

Chapter 3

INTERDISCIPLINARY APPROACHES TO MEDIA, TECHNIQUE AND ARTMAKING

On the first day of the semester in a high school Ceramics I class, the teacher asked her students, "What do you think we will do in this class this year?" The students replied with such answers as: "maybe make a vase for mom," "an ashtray for Uncle Bob," and other "things." One particular student, Neil[1] envisioned that he would make a set of dishes for his mom, along with some extracurricular "just for fun" objects. After his first week, Neil was surprised that his art teacher required more than just making objects with clay. His art teacher informed the class that they would learn about contemporary and historical ceramic works, ideas, techniques, vocabulary, types of clay, temperatures for firing ceramics, and health and safety, in addition to making "things."

Throughout the semester, Neil and the other students referred to maps, geography, time lines, and architecture. They also learned about cultures other than their own. After a couple of months Neil realized that some of the art that he studied in the Ceramics I class connected with what he was learning about in his social studies, history, chemistry, physics and English classes. For example, Neil learned that between 1979 and 1985 New York artist Joyce Kozloff designed and installed an 8' x 83' ceramic tile mural entitled *New England Decorative Arts* at Harvard Square Subway Station in Cambridge, Massachusetts (Cambridge Arts, Council, 2002, para. 4). He

Artmaking — Problematizing "Projects" — Ceramics
Skills — Inquiry and Research-based Approaches

examined images of the mural and discovered that Kozloff incorporated historical images and symbols related to the visual culture of the geographic region into the thematic sections. Further research of the art directed Neil to one such section of the mural entitled *Folk Art*. In this 8' x 17' section, Kozloff used 12" tiles that reminded Neil of his study of the "art" tiles produced during the Arts and Crafts Movement. He also discovered that Kozloff used quilt-like patterning and referenced gravestone urns, death heads, eighteenth-century engravings of clipper ships, a naïve New England landscape, Pennsylvania Dutch wall stencil designs, and wrought iron weathervanes. Beyond establishing numerous relationships with art technique and media—decorative arts, ceramics, stone carving, and metal sculpture—Neil, his art teacher, and the other students in his Ceramics I class used math skills to identify Kozloff's use of proportion. They referred to the practical engineering lessons in their physics classes to analyze the construction of Kozloff's architectural installation. They relied on the chemical properties they learned in their chemistry classes to understand the ceramic glazing process. And Neil referred to his social studies class as he discussed Kozloff's use of symbols to portray regional history.

Neil was, in essence, seeing and making connections between what he was studying and doing in his Ceramics I class with the ideas and ways of working in his other high school classes. As a result, he talked about what he was doing with his friends and wrote in his ceramics class journal about what the ideas meant to him. In the process, Neil's artmaking changed as he began to respond to these ideas in more personally meaningful and expressive ceramic works.

Connecting Ideas with Artmaking

In our experiences as teachers, pre-service educators, and observers in classrooms and other educational sites, the most exciting art making opportunities for students are those that provide a means to investigate past worlds in comparison with the realities of the contemporary world. Such opportunities are not limited to a single discipline, but rather are made possible through a wide variety of offerings in the school curriculum. If the disciplines in school are not connected, school is a disconnected assembly line of facts and assessments, where all the parts do not make up a whole education. **[See Chapter 1 for more discussion regarding disparate curricula in high school.]** Similarly, if the activities involved in the making of art in art classes are not connected to the real world experiences of students both in and outside of school, their interests in learning, critical and reflective thinking, and inquiry into cultural issues and ideas can be inhibited. "Individually, mathematics, science, and history convey only part of the reality of the world. Nor do the arts alone suffice. A multiplicity of symbol systems is required to provide a more comprehensive education" (Fowler, 1996, p. 47).

How can teachers design artmaking opportunities that help students correlate content among various disciplines? Like much of the art created by the greatest artists in historical and contemporary times, we believe that the art our students create should be reflective of an in-depth exploration of ideas as well as a mastery of skill, technique and media.

Essentially Artmaking is Interdisciplinary

Education practitioner and theorist Charles Fowler (1996) reminded us that the arts in schools do not, cannot, and should not exist in isolation. We cannot run from the fact that interdisciplinary connections are what we should do naturally in art education (Ballengee-Morris and Taylor, 2005). "Utilizing interdisciplinary instruction, integrating the curriculum and connecting issues and disciplines to student's lives stimulate relevant and meaningful learning while providing coherence for curriculum" (p.

Many artists combine their art making with other personal interests and ultimately may provide excellent interdisciplinary content and context for high school art and artmaking units of instruction.

13). We would add that interdisciplinary correlations naturally inspire relevant and meaningful artmaking. Just like artists of the past, contemporary artists use the disciplines of life for creative inspiration and interpretive import for their paintings and sculptures. Many artists combine their art making with other personal interests and ultimately may provide excellent interdisciplinary content and context for high school art and artmaking units of instruction.

By way of example, let us imagine an art class that is primarily composed of sophomores who are required to take biology during the same semester they are taking art. The art teacher might consider an assignment in which her students study the work of an artist who employs biological mechanisms such as Kendall Buster. "Sculptor Kendall Buster was trained as a microbiologist and explores the forms and landscapes seen in a microscope lens through her giant sculptures" (Ulaby, 2004, para. 1).

Figure 3.1 Kendall Buster, *Garden Snare (Shadehouse)* 1998, steel, green-house shadecloth, 11' x 19' x 9'. Kreeger Museum, Washington, D.C. Permission of the artist.

Just as any object—a work of art, visual culture, a piece of material culture—can be "read" using a variety of informational lenses (Sessions, 1998; Carpenter & Sessions, 2002, p. 373), the art teachers and students could approach Buster's work in several ways. They could read one of her sculptures through a thematic or idea-based lens as discussed in Chapter 2, by identifying the key concepts and essential questions of, for example, an image of her 1999 *Garden Snare* at the Kreeger Museum, Washington, D.C. (See Figure 3.1). Said to evoke a living and dividing cell (Ulaby, 2004), *Garden Snare (Shade House for Kreeger Garden)* is 11' x 19' x 9' steel frame covered in greenhouse shade cloth. To initiate research and further readings that attend to contextual, functional, and conceptual issues of the sculpture, the teacher could pose questions to her students, such as: What are the historical and cultural foundations of the sculpture? Does it serve a function? What philosophical ideas, and sociological-ideological issues are revealed? What are you reminded of when you look at the form and material(s)? How was it made? What does it resemble? Use what you see to guide you to the cues that are simply seen in the sculpture. What information from other school disciplines comes to mind based on the visual cues provided by the work?

The students and teacher might discover that sculptor Kendall Buster was inspired to create this work because of the microscopic images of red blood cells she witnessed when working as a lab technician. In fact, she often fantasized about inhabiting the tiny space only available for view through a microscope. Perhaps, the students and teacher will find images of the artist standing inside the sculpture as they explore ideas such as occupation or perspective associated with alternative spaces and viewpoints. **[See Chapter 1 for more discussion on interdisciplinary approaches to perspective.]**

The students could practice drawing microscopic images either by looking directly through microscopes in their biology lab or by studying digital microscope printouts. They could translate these patterns into their art work and/or look for patterns in works of art by other artists such as the cellular patterns found in the natural forms of wood-turning artist Brenda Stein (See Figure 3.2). The teacher and students could explore the dichotomy of interior and exterior spaces as they read that Kendall Buster's sculptures "exist simultaneously as object and as architecture, revealing the interchange of interior and exterior space" (Wendl, 2004, para. 1). The visual and conceptual cues they glean from the image of the sculpture may act as starting points for such research projects as an exploration of both biological and architectural scale, boundaries, and margins. Students may then consider the research, ideas, and meanings in the creation of their own works of art. They might be inspired to cre-

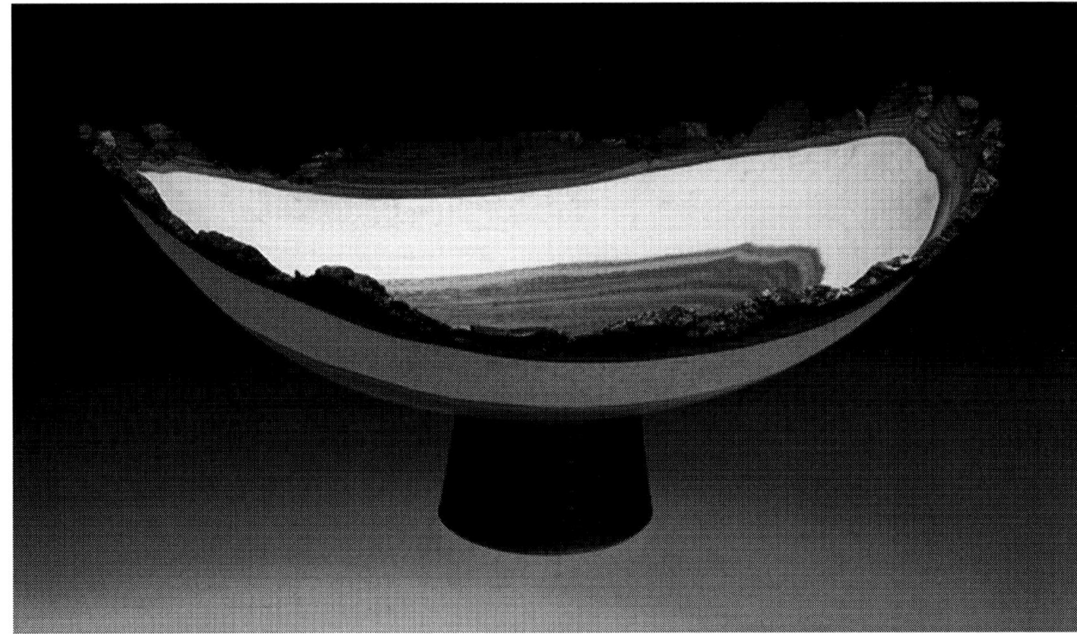

Figure 3.2. Brenda Stein, *The Shaman*. Turned wood. Permission of the artist.

ate models of exterior sculptural installations that like Buster's sculptural forms, challenge the viewers' conceptions of space.

The students could also be challenged to explore their own personal space and perspective by representing a corner of their locker, art tray, book bag or closet as a distinct and separate universe using sharp angled linear perspective or a dramatic three-point perspective technique. They might even compare Buster's sculptural inspirations and reflections of scale with Ibram Lassaw's 1960s projection paintings on slides[2] and digitally project their own small-scale ceramic sculptures onto alternative surfaces.[3]

The point is that teachers should enact ways of thinking and working that help students correlate their study with their own artmaking. We contend that such a connective artmaking process is quite naturally interdisciplinary. In other words, student artists instinctively integrate varied ways of knowing in their creative process when they are allowed and encouraged to do so. Caution must be taken to move beyond narrow formalist art teaching approaches that thwart such natural connections in the name of paying homage to the universal elements of art and principles of design. While all art is not the same and cannot be viewed or judged in the same way it is, however, naturally and meaningfully connective in one way or another. [See Chapter 6 for a further discussion on multiple aesthetics.]

Figure 3.4. Claudia Lee, *Paper Bowls*, 11" x 10" (2005). Permission of the artist. These bowls are made with handmade paper and created especially for an Empty Bowls project. (See Chapter 7).

The Interdisciplinary Nature of Pottery

Many cultures have a ceramics tradition that is rooted in the disciplines of living life. Archeologists who study ancient cultures typically refer to shards of pottery as their primary resource. As they do so, these researchers recognize the lengthy lines of interconnectedness between pottery and history. Ceramic objects are also among thc most significant artifacts that anthropologists study in order to understand cultures of the past. Potters and clay sculptors from the past and present, by their choice of materials, forms, skills, fabrication techniques, and decorations reflect multiple characteristics of their and other cultures.

Emphasis on the production of clay objects and the instruction of technical skills in many high school art and ceramic programs may have overshadowed the rich variety of connections between clay artworks and knowledge about the world. Ceramics for the sake of ceramics contradicts the origins of this craft-based art form, which was originally supported by the life of a culture.

In many cultures, food customs dictate the forms of objects while geography determines the types of clays and glazes that are used. In turn, the forms and formal properties of the ceramics objects help cultures to establish their customs and traditions. Differences in kinds of clays, the properties of minerals in glazes, types of kilns, kiln building, firing processes and fuels are interconnected to physics, engineering, and chemistry. Biology, social studies and ceramics are linked when scientists, archeologists, and medical doctors study clay figurines from pre-Columbian Central American cultures and ancient Egypt to determine specific diseases that afflicted the people. Some historical clay objects, writing instruments and accounting tools were closely related to the social and political development of societies and are therefore connected to economics and civics. For instance, porcelain was an important economic commodity in Europe because it was only produced in China and remained a well-kept secret for centuries. The finest porcelain objects made in China were so highly prized in Europe that they were used as money by the 12th century. Trade routes were established around the production and migration of porcelain objects. Ceramic tiles made in various countries and islands around the Mediterranean Sea, such as Italy, Morocco, Spain, Tunisia and Turkey reflect cultural vestiges as witnessed in their designs, colors, and glazing techniques.[4] (Charleston, 1976; Kingery, 1984-2000).

Careful examinations of the methods of construction, decoration and firing of ceramic objects may lead art students on various journeys into and outside of the world of ceramics and other school disciplines. A ceramic platter created by Peter Voulkos (Clark & Hughto, 1979) in his giant altered abstract expressionist style would direct students to a different learning space than the Joyce Kozloff subway mural discussed earlier in this chapter. Similarly, a Warren McKenzie tea bowl involves different technical and cultural experiences than a Maria Martinez black-on-black pit fired bowl (Sessions, 1999, p. 10). Different types of ceramic objects selected throughout the school year may take students on different thinking, making, technical, and educational expeditions.[5] A press mold ceramic vase with inlaid glaze inspires a very different learning journey than does a figure made of thrown and altered stacked cylinders. A slab stoneware tray with monochromatic Asian brushstrokes will guide students on a very different investigative journey than would a thrown

Figure 3.3. Billie Sessions, *Fragile Bodies*, (6" to 22") 2004. These stoneware and air-brushed glazed bottles represent her lifetime of "women's work." Using only one tool, or making do with as little as possible, she translates sewing techniques, with jewelry making and weaving processes. Hand-thrown thin slabs are stretched well beyond reason, demonstrating the fragility of an individual or life itself. Clay sheets are wrapped around kitchen and household cylinders as temporary molds, then pulled out when the clay is plastic enough to be released, yet almost firm enough to support itself—like children leaving home or students graduating. Granted, like people, there is a significant failure rate with her ceramic pieces. Within the repetitive nature of the individual shapes, each one presents new problems—as does parenting and teaching. Some are more straight and firm, some ragged and loose, some are intentionally misshaped, some unintentional, and some are well groomed and full of themselves, yet hollow.

earthenware, polychrome majolica sugar and creamer with Spanish ornamentation. These alternative journeys could begin with such questions as: To what ideas does the ceramics object visually refer? What was the artist's objective? To what realms of information does this ceramics bowl connect?

Glacial Flower, an irregular bowl by Sally Resnick Rockriver (Rockriver, 2004), has deep throw rings left in the bottom and cave–like protrusions with thick glaze dripping that puddles in tiny pools. Handmade protrusions and tiny reservoirs curl around the rim of the bowl and resemble calcium formed stalactites, geodes, and tide pools. The subconscious geologic and fantasy worlds referred to by *Glacial Flower* could be compared to a Harry Potter[6] landscape.

Many contemporary clay artists are concerned about art for society's sake, rather than art (ceramics) for art's (ceramic's) sake. (See Figure 3.3). Elaine Levin's 1988 survey text of American ceramics illustrates the following connections. The Yixing teapot series of Richard Notkin comments on nuclear war, trash sites, and the structure of the human heart. The work of Patti Warashina refers frequently to women's issues through giant figures of mythological women or diminutive females in all sorts of spirited domestic scenarios. Adrian Saxe makes ceramic forms that conjure associations with French royalty, geology and household tools on one hand and issues of social and moral values on another. Blatantly, Robert Arneson made fun of historic and contemporary politicians, and the darkness of war in his realistic sculptures of people and human-made objects and environments.[7]

Some ceramic artists create work that blurs the lines between disciplines. For example, artist Buster Simpson related art, science, and activism to social issues in his ceramic plates (Cembalest, 1991). His three-part partitioned ceramic plates imitated heavy paper picnic plates. Simpson placed the plates where water from sewage treatment systems flowed over them for several months in Puget Sound, Love Canal and New York's East River. The so-called "clean" water should have allowed the plates to remain white, however, the water from the treatment systems left ample toxins that fully glazed Simpson's ceramic plates.

How do we make art that reflects our ideas and concerns? Kathleen Hall, art teacher at A. C. David High School in Yakima, Washington requires her ceramics students to use personal symbols, signs and designs to create a ceramic storyteller representing a memorable event in their own lives. **[Teacher contribution.]** When art students and teachers look closely at work by these and other artists in conjunction with or in preparation for making their own art they may ask: What separates works of art and art making from research and activism? What relationship can works of art have with research and data collection about pollution and other environmental issues? What do we need to know in order to create meaningful works of art? How do works of art, their creation, and their functions teach us about the world and ourselves?

Another artist whose work and artistic practice raise questions about distinctions between art and other disciplines is Joan Lederman, a self proclaimed "potter and imager" (Kemp, 2000, p. 134). Lederman uses sediment from the floor of the Atlantic and Indian Oceans from 4500 meters below sea level as glazes on her ceramic vessels. The sediment she uses is 35-40 million years old. "To cradle one of her

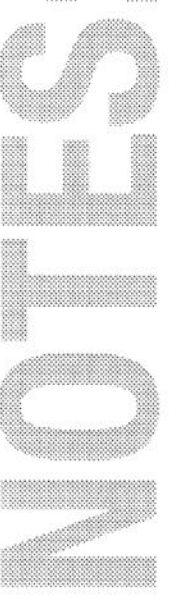

There is a difference between what we call art and what we call projects that result from exercises with media and techniques.

vessels in our hands is an evocative, even eerie experience, akin to holding a fragment of meteorite or peering in wonder at a sample of moon rock" (Kemp, 2000, p. 135). How do materials and media affect the meaning of a work of art? Using Lederman as an example, one possible answer is that "... she is actively inducing us to share her sense of wonder with the way that the earth, fire, and water – the very stuff of the potter's art – speak of elemental forces that have repeatedly attracted human creators across distant times and spaces" (Kemp, 2000, p. 135). But is that the only answer to this question? And is that the only question that works like Lederman's suggest? For example, how do scientists know that the sediments are 35-40 million years old? And why is it important to know how old the materials are? What do moon rocks, meteorites, and space dust tell us about life on Earth or other parts of the universe? Why did certain historical cultures develop certain pots, glazes, and firing techniques? In what ways does geography and climate influence the creation and use of artifacts in a society?

These few examples are rich points of departure from which students may begin to make interdisciplinary links to historical and contemporary issues. Works such as these are also powerful starting points for curriculum designers who are interested in creating interdisciplinary art and ceramics curricula. Although we have not deeply explored the connections suggested by the works of these artists here, we have demonstrated some of the ways that ceramic works of art may connect to nuclear power, circulatory physiology, city landfills, ecology, mythology, politics, African tundra, French aristocracy, classifications of rocks, industrial revolution and tropical botany. Again, we believe that authentic interdisciplinary connections should also include important personal connections that are relevant and meaningful to both the teacher and her students. Therefore, we offer these examples as rich starting points from which high school art classes can make meaningful connections.

Moving Beyond Media and Technique-based Instruction

Before we go on, we must recognize the problematic nature of our argument against media or technique only based instruction in art education through our example of a class that is actually titled according to the media and technique—Ceramics. Similarly, when an art course is titled painting, drawing, photography, or sculpture, it obviously implies an exploration of a technique or media. Although other class titles—Art I, Art II, Art III, and Art IV—may appear less problematic to this discussion, we actually question whether an authentic interdisciplinary approach to high school would involve specific courses much less specifically titled ones. Perhaps mini-courses, experiences, or workshops that are based on ideas would be more in line with this discussion. Or maybe as Theodore Sizer (1993) recommended, we need to completely restructure the high school system to combat the fragmentation suggested by disparate courses and curriculum.

We realize such thinking may be overwhelming at this juncture. And since the purposes of this book are to inform and assist pre-service and practicing high school art teachers to infuse interdisciplinary and integrated ways of thinking, teaching and learning into existing high school systems, we will refrain from postulating further. We share these concerns in an effort to further emphasize that authentic interdisciplinary teaching is beyond complicated. It is indeed naturally complex.

In fact, all teaching and learning can be complicated or complex and such systems help us determine the most effective instructional approaches for interdisciplinary content. Complicated systems are based on mechanical metaphors of organization, such as grids, rubrics, and clocks. Such complicated systems are "human-produced mechanical systems" and are "the predictable sums of their parts. Their behaviors are planned, directed, and determined by their architectures." (Davis, Sumara & Luce-Kapler, 2000, p. 55). Complex systems are more alive, organic and self organized like conversations, vines and weather systems. They "exceed their components" and "Unlike complicated (mechanical) systems, which are constructed with particular purposes in mind, complex systems are self-organizing, self-maintaining, dynamic, and adaptive." (Davis, Sumara, and Luce-Kapler, 2000, p. 55).[8] Visual art is not an isolated discipline, and sadly it becomes ineffective when art teachers base their lesson planning solely on media in a complicated approach, rather than as a vehicle for creating meaning through a complex one.

In other words, rather than basing curriculum on learning a medium or technique, we recommend that art teachers base their units of instruction on a problem, issue, or question gleaned from works of art and visual culture and challenge students to solve these problems and/or address these issues through an exploration and mastery of media and technique. In our minds, teachers should replace a media and technique-based approach with problems, issues, or questions from which students must select the media that they believe best suit their responses. Granted, students will need instruction in some media and techniques, but such instruction is given in the form of exercises much like the practice of scale in music class, verb conjugation in Spanish class, and sentence construction in English composition class. In piano music, scales are an essential resource. They assist in learning and mastering fingering techniques that assist the pianist to focus on expressive melodic contour (Osterland, 2005). Similarly, mastering Spanish verb conjugation and sentence construction plays an important role in communication and expression. In other words, these exercises assist in a mastery of technique that ultimately serve as methods or ways of working to express meaning. There is a difference between what we call art and what we call projects that result from exercises with media and techniques.

In our minds, teachers should replace a media and technique-based approach with ideas, problems, issues, or questions from which students must select the media that they believe best suit their responses.

Neil, the art student we discussed at the beginning of the chapter, was challenged by his teacher to create a work of art that reflected his personal association and/or value of the geographical region and community in which he lived. He wanted to draw from the tile construction of Joyce Kozloff's *New England Decorative Arts* (Johnston, 1985), but wanted to add more three-dimensional effects and therefore needed to master certain ceramic skills and techniques. With the guidance of his

Photograph by Donna Green, 2006.

teacher, Neil practiced slab rolling, slip applications, and additive and subtractive modeling techniques. He experimented with different glazing techniques while at the same time practiced drawing with proportion and perspective. His final work of art was a 12" ceramic slab cylinder covered with tiles on which he painted local landscape scenes with underglazes. Scattered about the tile-covered cylinder were intricately modeled figures of elk, deer and moose as well as other architectural and geographical landmarks. Neil wanted to create a work of art that reflected his deep and passionate association with the grand and exquisite vistas and wildlife of the western Wyoming community where he lived. The exercises in which he engaged with media and technique, like piano scale exercises and Spanish verb conjugation, contributed to his ability to create a work of art that expressed what was meaningful to him.

Through an issue, question, or problem-based approach, students become creative problem-solvers who also learn the skills and processes necessary to create works of art. That is, students will use the media that best serve them as they develop their responses and ideas. In this approach, students learn techniques and media to voice and embody their ideas. If we are to move beyond the limitations of the twentieth century elements of art and principles of design as the basis for art curriculum, then we need to transform the media and skills based paradigm to more authentically represent what they are— methods toward meaning.

On the other hand, now more than ever contemporary artists create works that examine and question the contemporary human condition by amalgamating media. In fact, contemporary artists frequently use non-traditional materials because of the symbolic meanings and references those materials carry. (See Figure 3.4). As extensions of the modernist paradigm yet in the tradition of Salvador Dali and Robert Rauschenberg, contemporary artists Mierle Laderman Ukeles and Lynn Hull (Matilsky, 1992) use salvaged materials instead of paint, charcoal, and clay to explore links between daily life and environmental issues. Some university art departments have even gone so far as to knock down the literal and figurative boundaries that have divided the studio areas in their programs. In response, students have cross-fertilized their ideas through the media that work best to convey their message. **[See a discussion of flexible purposing in Chapter 10.]**

Let us now look at an example of a media non-specific art course where students were required to create an assemblage based on a current social or cultural issue in their community.[9] One student selected pollution as his focus. He lived in Virginia Beach, Virginia near Mt. Trashmore, a landfill that had recently been turned into a park. The park had a lake, benches, and grass. He and other local residents were hesitant about swimming in the water and many would not visit the park as they feared that toxins in the ground would seep into the grass under their feet and ultimately into their own human systems. For this assignment, the student built a 1.5" x 2" wooden shadowbox structure with a Plexiglass™ front. He filled the box with trash—soda cans, food wrappers, paper and metal scraps, foam cups, plastic debris—and placed a layer of sod on the top. The grass was exposed to the open air. Viewers of the work could see the layers of trash through the clear plastic. Over the period of days in which the work was exhibited in the classroom the student watered the sod and the grass continued to grow. Through the media and the performative nature of the work, this assemblage inspired numerous questions from classmates about the health risks of walking on grass that grows on top of trash or landfills and eating fruits and vegetables produced in similar conditions. In the class discussion the students agreed that this artform and the media used to create the work were more appropriate and powerful choices to communicate the student artist's perspective on the environmental issues inspired by Mt. Trashmore than a traditional painting or drawing.

Projects, Problems, Inquiry, and Research

Art teachers often refer to the work their students' create as projects. We challenge this concept and the use of the term "project" especially at the high school level. We know that some pundits say that student work is not art. And we have witnessed students who stubbornly did not want to work on a particular drawing or painting any longer, refused to respond to the teacher's suggestions, and declared, "this is my art and I say it is done." Our response to situations such as this is, "Yes, this is your art and you should make your own decisions about what and how you create it. However, in addition to creating something visual (form), the point of an art work is also to convey meaning (content) and as a student you are here to learn and practice ideas and techniques for doing both."

As we explained in Chapter 1, a project is a planned undertaking, task or problem in which students are engaged to supplement and apply their classroom studies. Projects are activities that assist students in developing skills and ways of working that lead ultimately to their understanding and making of meaningful idea-filled works of art. As many art teachers will attest, hands-on creation requires students to think inventively and to utilize a variety of skills they learn in and outside the art class.

We believe that visual art is a body of knowledge worthy of study and that the creation of works of art with symbolic content requires divergent, inventive, and meaningful thought. Such an approach and mindset moves beyond traditional ways of conceiving of "projects" as simply objects or images that students make in art class. In our view, student works of art are complex constructions that embody and reflect meaning and learning that students derive from a combination of responsible and engaged thought, visual attention, critical reflection, and purposeful artisanship.

Project-based interdisciplinary learning in and through the study of art can also include projects that do not lead to artmaking. For example, art students might create a report, a garden, a fundraising initiative, or a performance. Projects can begin with a single discipline and build upon or reference knowledge and skills from other disciplines along the way. Interdisciplinary instruction projects might also be service oriented, such as cleaning a park, landscaping a school garden, conducting after school tutoring and reading sessions, or helping at homeless shelters. [See Chapter 7.] That is, an interdisciplinary project need not be an object or artifact but might be a performance, experience, or on-going event. Regardless of whether an artifact is produced as a result of the project, learners become intellectually engaged when the projects they are working on are personally meaningful (Kafai and Resnick, 1996). (See Figure 3.5).

> Many art teachers already consider problem-solving to be the essence of any and all art making experiences. Whether the problem is technique, media, or concept-based, artists and art students constantly work to find solutions in their creative processes. When teachers involve students in creative problem solving, they also invite their students to participate as partners in the learning process. Instead of telling them what to think, these approaches to learning in the arts engage the minds of students to sort out their own reactions and then articulate them through the medium at hand. Students become embedded in the task so that they learn from the inside out rather than from the outside in (Fowler, 1996, p. 48).

Making meaningful correlations among and between various disciplines is essential when teachers intentionally encourage students to combine content to solve problems. For example, Stacy Potter, a ceramics and sculpture teacher in Mesa, Arizona teaches a design lesson in which her students use computer technology and geometry to construct elaborate patterns that relate to historical quilts. **[Teacher contribution.]** In such a problem-based

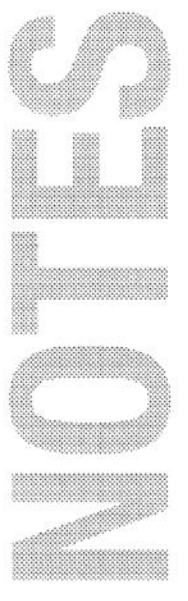

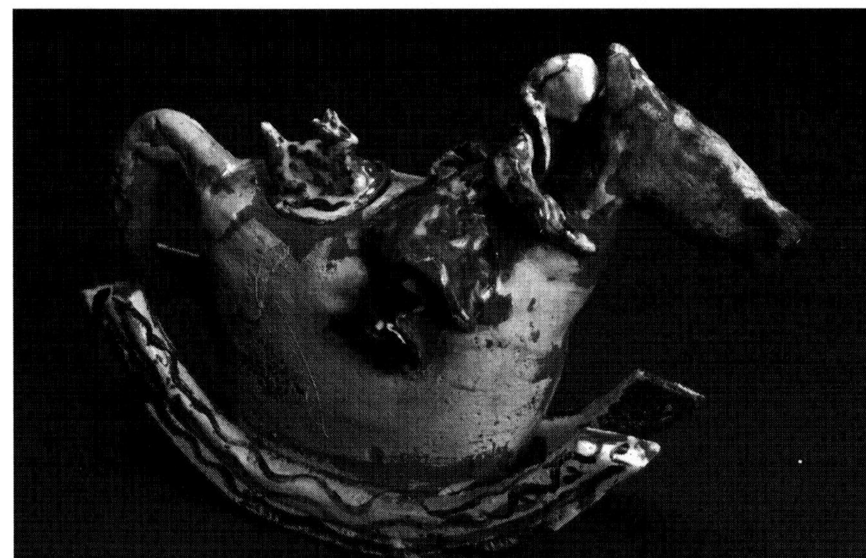

Figure 3.5. Donna Rizzo, *True Companions*, 12 x 12 x 4 inches ceramic teapot. (2005) Permission of the artist.

assignment, her students learn more than simply how to use a computer to create patterns more quickly than they could by hand. This problem-based assignment requires students to consider mathematical questions such as probability, ratio, and balance. Before Mrs. Potter's students can render the patterns they design they must experiment first or play with the computer program to determine the limitations of the machine with respect to rendering images and then redesign the symbols that will appear on their ceramic quilt blocks accordingly. And when students create a contemporary quilt with clay and technology based on their understandings of historical quilts, their teachers should not limit them to "traditional" looking quilts. That is, why would a contemporary computer-generated or ceramic quilt need to look like a 17th or 18th century fabric quilt? In this problem-based assignment, students should question the ways in which technology – looms, needles, sewing machines, computers, laser printers, photography, and fabrics– influence the manufacture, function, size, and meanings of quilts, past and present.

Problem-based approaches to teaching and learning in art do not always result in student artmaking. What would a non-artwork problem look like? Although not an art educator, we suggest that teachers take the lead from Murray Burton Levin (2001) who encouraged teachers to:

> Create lesson plans that require students to locate an incongruous or "deviant case:" fictional characters, for example, who suddenly behave profoundly out of character or variance that produce results very different from those intended. The challenge for students is to identify the unpredictable by locating the idiosyncratic forces. This, of course, is one of the first steps in science, and an exercise that sharpens logic. (p. 153).

Inquiry and research-based instruction requires teachers to assist students to develop strategies that seek meaningful answers to important questions. Teachers must help students understand that the first or easiest answer might not always be the best or only possible answer. Through inquiry and research, students, teachers and scholars uncover important meanings about people, cultures, and societies. Most significantly, in an inquiry, problem or research-based approach,

> Teachers must make it absolutely clear to students that no issue is really settled, no answer is really sufficient, unless students can explain why the phenomenon occurred. Teachers must impress upon students the subtlety of these issues by demonstrating how definitions determine findings. The salient question is always why [italics in the original]. (Levin, 2001, p. 152).

Because most forms of inquiry and research occur experientially, we believe that teachers should help students study the world by asking "*why?*" and experiencing it first hand (Fowler, 1996). Such approaches to inquiry and research might require students to react or respond in a variety of ways.

> The arts – creative writing, dance, music, theatre/film, and visual arts – serve as ways that we react to, record, and share our impressions of the world. Students can be asked to set forth their own interpretation of the Grand Canyon, using, say, poetry as a communicative vehicle. Whereas mathematics gives us precise quantitative measures of magnitude, poetry explores our disparate personal reactions. Both views are valid. Both contribute to understanding. Together they prescribe a larger overall conception of, in this case, one of nature's masterpieces. (Fowler, 1996, p. 47).

Along these lines, students can create brief digital video mini-documentaries or mock television news stories in which they interview experts about a question, issue, or situation. These experts could be people in their community who have first-hand knowledge about the situation, such as district rezoning, segregation, a hazardous traffic intersection, racism, sexism, or unfair labor practices. But simply documenting the perspectives of these experts is not enough to constitute a sufficient documentary in response to an inquiry or research-based assignment. Student mini-documentaries or mock news stories would also have to include text, images, and voice-overs that describe the situation, demonstrate evidence of research using primary and secondary sources, and explore more than one side of the story. Through the availability and ease of digital video editing on desktop and laptop computers, students can easily edit their documentaries and news stories and publish them on DVD or online. If carefully structured around well-conceived problems, these student works could satisfy characteristics of project, problem, and inquiry/research-based instruction all at once. **[For more on digital video and visual culture, see Chapter 8 and 9.]**

Conclusion

A wide variety of interdisciplinary approaches are available to teachers who consciously merge the content of various school disciplines in the curricula they design. These project, problem, inquiry and research-based approaches to instruction, over time, enable students to learn to see the power of thought and how it can be used to promote change, knowledge, and understanding (Levin, 2001). The bottom line is that such interdisciplinary approaches help students correlate the curriculum with their lives in meaningful ways.

The result of such ways of working extends to not only what we study in our interdisciplinary high school art classroom, but also in how we and our students make art. Art created by students becomes much like the great historical and contemporary art that students study when they correlate their understandings of the interdisciplinary and integrated powers of art with their own artmaking. Their reflection of an in-depth exploration of ideas joined with a mastery of skill, technique and media, becomes a work of art that is a personal and social expression of meaning.

Suggested Discussion Questions and Activities

1. At the beginning of this chapter, Neil discovered that his art teacher required much more of the class than he originally thought. One important dimension of the class that was not discussed in this chapter involved health and safety in the art classroom. Read Safety in the Art Room published by the National Art Education Association (http://www.naea-reston.org/classroom.html). From this reading and further research answer the following questions: What is a material safety data sheet and why is it important to have a file with MSDS sheets of all supplies and materials in your art classroom? What steps should art teachers take to

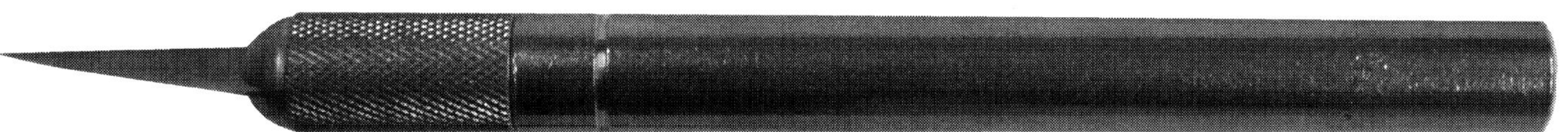

Photograph by Donna Green, 2006.

protect their students as well as themselves? What safety plan is in effect in your school? What would you do if you experienced a chemical spill in your class room? What would you do if someone were bleeding? What other important health and safety procedures are necessary in the art classroom?

2. Some students come to art class in high school with predetermined ideas about what they will do, make, and study. How will you respond to such student expectations that contradict your plans for nontraditional and problem-based instruction? different possibilities for art in life, high expectations right off the start, grading criteria you might not succeed at every-thing but you can find something you really love and use that to your advantage

3. Greenmuseum (http://www.greenmuseum.org) is an online museum of environmental art. This site has information for educators and communities who are interested in pursuing environmental projects. Visit this site and select an artist whose work you find compelling in some way. Design a set of three learning activities for high school students based on the work, media, and key concept of the artist.

4. Numerous curriculum examples exist online for teachers interested in interdisciplinary approaches to teach art and science concepts. Visit the Art in Science Project curriculum (http://www.odu.edu/al/artinscience/completecurriculum.htm) and design a learning activity based on the photomicrographs on this site.

Notes

[1] Neil is a fictional blend of students in author Billie Sessions' former high school classes.

[2] "The paintings themselves were made on tiny glass slides in the 1940s, but shortly before his death in 2003 Mr. Lassaw supervised the transfer of some of the images, which look as rich and resonant as they did more than 50 years ago" (Harrison, 2004, para. 1).

[3] "Through grand scale audio-video projects in public spaces, artist Krzysztof Wodiczko transforms national monuments and architectural façades into "bodies" as he collaborates with communities to get people to "break the code of silence, to open up and speak about what's unspeakable" (PBS, 2005, para. 6).

[4] A prime example is the Islamic 13th century architecture and tile at the Alhambra fortress in Granada, Spain. A highly desired commercial bastion of the Mediterranean, the island of Maiolica (later named Majorca, now Mallorca) off the coast of Spain was the primary departure port for Spanish-Muslim earthen white ware well before the 11th century. The early wares often made of red earthenware clay were coated with lead or tin-based white glaze to mimic the highly prized Chinese porcelain. Though they most often featured blue (cobalt) designs, the white glaze was soon topped off in various design styles and color(s). These wares were then sent to destinations in Europe and Asia Minor. Because it was shipped from the island of Majolica, the earthenware that was sent to Italy came to be known as maiolica or majolica. Later on the white ware made in Italy (called Faenze) fashioned after ware from Mallorca became highly popular in 16th century France and came to be known as Faience. The same style of ware is called Fayence in Germany, and called Delft in Holland (The Netherlands), the primary production city. Delft is most often white with cobalt blue much like the early Chinese porcelain designs.

[5] This holds true for any artwork or cultural artifact. Paintings by different artists, from different eras or subjects will evoke different educational journeys. Different kinds of printmaking, photography, sculpture, and contemporary and historic cultural artifacts serving different purposes from various points in time will lead the students to different processes, ideas, and interdisciplinary connections.

[6] See htpp://harrypotter.warnerbros.com

[7] See also Kangas, 1980; Lewis, 1991; Lynn, 1990

[8] Traditional approaches to teaching and learning, such as behaviorism, are built on theories similar to complicated systems. In contrast, holistic and child-centered approaches such as constructivism shift the focus of learning and offer strategies not possible with traditional approaches.

[9] This story is based on author B. Stephen Carpenter's recollections of the work of a student in a class he taught at Old Dominion University in 2000.

Chapter 4

TEACHING MEANS WE'RE MEMBERS OF A TEAM

In August Ms. Walker,[1] the art teacher was transferred to a new school. The principal assigned her to a team. Each team consisted of "core teachers" and "encore teachers." "Core" referred to teachers of academic areas such as math, English, social studies. "Encore" teachers taught other subjects like art, music, and physical education. Already Ms. Walker felt different, not the same as the other teachers in the school. The core teachers internalized that encore was extra and not really educational. In the team meetings, Ms. Walker offered ideas, but felt invisible. The core teachers teased her kindly but nonetheless their words were hurtful. "We know you mean well, but we teach students meaningful information and you teach how to draw stick people—like that will change the world or get our students a job." Ms. Walker wanted to respond, but did not know how. So she went along and tried to connect her class plans to the themes that the core teachers had chosen. One day in the hall, the English teachers saw Ms. Walker carrying books to her classroom and seemed genuinely surprised. And to top it off, the computer technology teacher chose to welcome Ms. Walker to the team by placing a banner outside of the art classroom, "Welcome to the Cut and Paste Department." Did Ms. Walker feel like a member of the team? How much power did she give the other teachers? What part of the responsibility was hers to educate the other teachers in the school? What strategies should have been created to prevent or change this scenario? When one is hired as an art teacher, one automatically becomes part of a team—all teachers at a particular school are members of a team whose job is to educate students from a certain community. Depending on the culture and support of a particular school, teams can

Motivation — Advocacy — Staff Development

Collaboration and Voice — Team-Building

be dysfunctional or functional. When team members label and/or place hierarchical value on each other as in Ms. Walker's experience, the team is dysfunctional—and makes students' learning outcomes problematic at best.

We titled this chapter associate because we define "team" in a deeper way than simply meeting as a group once a day, week or month. So what does associate mean? Why use that term associate when categorizing team teaching or team building? Associate can be a noun or a verb. When people associate there is a mental and psychological connection from one to another. This connection is developed through spending time together. Through that process we form relationships and often these relationships become associations. Even in the best teams, hierarchies can emerge, tendencies toward individualism can persist, and genuine agreement can be elusive when not approached associatively. An associative approach to team building should remove notions of competition, hierarchy, resistance, and taking sides and replace them with shared and collaborative control leading to professional inquiry and curricular and pedagogical innovation.

In educational settings, team teaching is defined as an instructional program that involves two or more teachers who teach two or more subjects in a coordinated way. We recommend that equitable participatory approaches be integral components of team planning and teaching. Meaningful collaboration is the foundation for successful development and implementation of interdisciplinary/integrated curricula, teaching, and learning. We therefore advocate not only rethinking terms such as teams, accountability, and leadership but also viewing those ideas as a part of collaborative practices.

Collaboration is the act of many people working together to achieve consensual goals as associates. If everyone is an equal partner, voice, and leader, then each is also equally invested and accountable. Paulo Freire stated, "Art creates community, community creates coalition, and coalition demands change (as cited in Ballengee-Morris, 1998, p. 115)." Collaborative practices create community, which in turn create coalition for school reformation. Ms. Sara Brock, a high school teacher, stated that being a team member and working as a team was essential to interdisciplinary and integrated teaching and learning in high school. [Teacher contribution.] When teachers view each other with respect and as colleagues, they are able to rethink their field and how they can connect to each other for "our" students rather than "my" students. [Take a look at Chapter 5 for more on Paulo Freire and teachers and students as co-learners.]

Because interdisciplinary education generally falls outside the norms of the culture of the school or the school system, key individuals and groups must offer support to launch and sustain the program. Such support is both necessary and essential especially if the proposed undertaking is costly or complex. We recognize that a department head, principal, curriculum director, superintendent, school committee, or fellow teacher may hinder as well as contribute to an interdisciplinary education proposal. The key, we believe it be proactive and ready if and when resistance occurs.

When associates engage in collaborative practices to form community and coalition for reformation, they become change agents with genuine support and respect for each other. After proposing, designing, and teaching a course in Art & Literature, Ms. Sara Brock and Mr. Mark Graham asked these important questions: "How can high school teachers cross boundaries that have remained intact [for centuries]? How can they put aside the customary notions of specialization (on which they have built their whole careers)? How can they meet all the demands of their home department while making room for something vital and new in the interaction with another field?" (personal communication, April 22, 2005).

School reform requires structure. One way for teachers to begin the process of school reform is to establish a common language that goes beyond disciplines and thus breaks down barriers. **Collaboration is not only an action but also a metaphor and foundation of integration.** Methods of collaboration, the pitfalls and pluses, the processes, accountability, and releasing/renegotiating power are all subject to change and fluctuation.

Each of these aspects must be discussed initially among all members involved. It is important to revisit these aspects when necessary. Resistance to being a team and to team teaching is often found as a response to time constraints, "I don't have time to know the other subject areas." Initially, collaborative planning and designing interdisciplinary/integrated curricula are time consuming because much of the time is spent on how, what, and when. Once the structure, discipline, and system are developed and routinely utilized, the time factor should not be an issue.

Collaboration, if used correctly, is time efficient. The time is there — learning how to use it and how to develop the proper attitude requires small shifts in practice and conception from all teachers who are involved in the practice. In this process, focusing on the objectives of the program and not the by-products is essential. Viewing curricula as a process and not a product is the key.

Implementation is the initial stage, not the end. In fact, there is no end to the planning process because practice and theory inform each other. The collaboration process also requires that teachers establish assessment tools, revisit concepts, and develop a conceptual scope and sequence that reflects district and state requirements. The amount of themes, units, and activities that collaborative teams utilize are insignificant and are only by-products at most. We have found that in the most successful collaborations focus needs to be on the process, which becomes the product.

The investment teachers put into becoming a collaborative team is crucial to the foundation of integrating curricula. This investment comes in forms such as professional development, learning about each other, developing a climate of trust, and learning how to be risk takers. A collab-

Collaboration, if used correctly, is time efficient. The time is there—learning how to use it and how to develop the proper attitude requires small shifts in practice and conception from all teachers who are involved in the practice.

Photograph by Donna Green, 2006.

orative learning climate provides a place for reflective and critical analyses of influential factors in learning and knowing. In this process, it is important to continually reflect upon personal and communal goals. A collaborative learning climate promotes the celebration of successes and provides support and guidelines for enhancement.

Developing an Associate Voice

Finding one's voice is the first step in team building. This process helps everyone learn about the other people on the team and establishes equality in terms of power, respect, and importance. Such an approach eliminates the tendency to establish a hierarchy. Equality strengthens intergroup relations and helps to develop a climate of trust. Teachers who participate in teams often disregard the importance of investing time in this usually assumed or ignored step (Brewer & Miller, 1984; Du Bois & Hutson, 1997; Miller & Harrington, 1992). When teachers on a collaborative team take time to get to know themselves and each other, they decrease perceptions and assumptions, and instead recognize differences and commonalities. The process that such an approach takes will also depend on the culture and resources of the school.

In *Leadership Challenge*, Kouzes Posner (2002), stated that finding and developing one's voice is critical to becoming a leader (p. 44). Since a collaborative team requires everyone to be a leader, each person must find their own voice and leadership through discovering their own individual strengths. What are the values, beliefs and assumptions that drive you? Do you practice what you think?

Values are guidelines that influence our lives, responses to others, actions, and commitment to schools' goals. Posner (2002) observed that values are our personal "bottom line," which reflect why we are teachers, as well as our philosophies and practices (p. 48). Ask yourself: Why do I teach? What motivates me to be a teacher? What are my thoughts about learning? Do I consider myself still learning? What is it that I value? Do I value mentoring? Do I value learning about others? How do I reflect those values in my daily tasks, decisions,

During professional development workshops and summer institutes, teachers work individually and in small groups to expand their thinking about curriculum and instruction. Contemporary Art Center of Virginia, 2005.

Why do I teach? What motivates me to be a teacher? What are my thoughts about learning? Do I consider myself still learning? What is it that I value?

and teaching? Posner also recommended that teachers ask themselves about the books and stories that made the biggest impression on them as children. Then ask: What values did these books teach me? What books or stories am I reading right now? What values do they teach? All of these questions help teachers begin to develop a sense of self-awareness and understanding [See discussion question #4].

Teachers immerse themselves in arts experiences and learn how to integrate the visual and the performing arts across their curriculum at the 2006 Partners in the Arts' Teacher Training and Assessment Project (TTAP) Summer Institute at the University of Richmond in Richmond, VA.

Do I value mentoring? Do I value learning about others? How do I reflect those values in my daily tasks, decisions, and teaching?

Finding one's voice is also about discovering one's strengths and what one can offer to a team that goes beyond the discipline one teaches. Strengths are a combination of talents, knowledge, and skills. Marcus Buckingham and Donald Clifton (2001) wrote *Now, Discover Your Strengths*, a book based on 2 million interviews with corporate leaders. In the book the authors suggested that teachers should not focus on ways to improve their weaknesses but rather to understand and improve their strengths. Buckingham and Clifton also recommended that the second step in knowing oneself is to discover one's strengths and to list them. For example, if one's strength is to work independently, how can one use that strength in a collaborative team? In one school, Ms. Debon[2] resisted the idea of being on a team and often stayed silent during team meetings. She usually stated that she was not a "joiner" and preferred to work alone. At one team meeting, one teacher remarked how impressed she was with Ms. Debon's technology abilities and referred to the new website she had created. As the meeting progressed, the teachers expressed their desire to document their project for the community to view. At that moment, Ms. Debon volunteered to develop a website. It was her first step toward becoming an active member of the team. At first, this outcome may seem like a small change and a reward to Ms. Debon to further isolate herself. However, this move to join the team incorporated Ms. Debon's strengths and was self-driven, a characteristic that helps to decrease, if not limit resistance.

Once someone has found their strengths and leadership qualities, the next step is to share those qualities with others. When people share their strengths and leadership qualities they also help to identify similarities between members of a team and illustrate potential issues or voids that could be addressed in a proactive manner. Similarity in values, practices, or history, leads to attraction and helps de-emphasize differences that are related to increased bias (Osbeck, Moghaddam, & Paerreault, 1997). As each member recognizes, acknowledges, and appreciates each other's unique and specific cultures, languages, histories, strengths, values, and subject areas, the process of building a team and maximizing each other's strengths when working together begins.

When members of a collaborative team identify potential conflicts before a disagreement occurs and emotions are invested, they create opportunities to improve abilities and to achieve common goals. Ms. Nan and Mr. George were both teachers[3] and members of the same high school history team. Ms. Nan was a respected history teacher and an avid fan of a professional sports team that used Native American people as a mascot. Often Ms. Nan came to work dressed in apparel from that team. Mr. George on the other hand, was an American Indian who taught political science. He was a member of a national group that often demonstrated at the stadium where Ms. Nan's favorite athletic team played. When the school team met, all of the members felt tension in the room. One day Mr. George brought in an article published in *Teaching Tolerance*,[4] that addressed mascot issues and American Indians. He gave a copy of the article to everyone on the team. As a result, Ms. Nan and Mr. George were able to discuss the issues at hand. Although Ms. Nan continued to be an avid fan, she became more aware of the deeper issues and worked with Mr. George to explore those issues with their students in a team teaching project that encouraged negotiation and social change.

The key to the resolution of the tension and conflict about the American Indian mascot imagery is that Mr. George gave a copy of the article to everyone on the team. He did not single out Ms. Nan, but rather his focus was on a broader sense of understanding and resolution. When collaborative team members make mistakes or have different views and are forgiven, the focus is most often on how to negotiate effectively in a way that considers all parties. Conflict transformation challenges the process and addresses the development of a sustainable, constructive, just, and equitable relationship among group members.

When people share their strengths and leadership qualities they also help to identify similarities between members of a team and illustrate potential issues or voids that could be addressed in a proactive manner.

Associate Support

At this juncture, we believe that it is important to note that some types of resistance can be viewed as conflict, strength, or weakness. Statements such as "I don't like change" or "I don't do well with change" or "it takes me a while to get used to things" indicate forms of resistance. In fact, most people do not like change unless it is their idea. We mention this now because often resistance is listed as one of the obstacles in integrated curricula development, team development, and school reformation. Knowledge of the possible forms resistance may take helps other team members to understand why resistance is happening, what should be listened to, what should be addressed, and ways in which the resistors can be supported. Teachers can be proactive when they offer opportunities for resistance to be valued, analyzed and reformed into positive actions.

> If society wants to help students grow into independent and knowledgeable thinkers who are capable of democratic deliberation, society must expect teachers to be similarly capable—and seen as such by students. To create an effective culture of democratic professionalism, school leaders must expect and model professional discourse, including the establishment of norms of critical feedback and collegiality (Gunn and King, 2003. p. 195).

Associates with Shared Visions

Developing relationships within and between groups of people is most effective when establishing goals that recognize and celebrate individual strengths. Collaborative learning communities that exhibit democratic deliberation, inquiry, trust, and risk-taking among teachers are important to a school's capacity for instructional change and lead to high and equitable student learning. Learning about and respecting others creates an atmosphere that promotes coalition and community development.

In the story at the beginning of this chapter, the teachers assumed the art teacher taught students how to draw stick people and primarily worked with cutting and pasting. These teachers had little knowledge about the field of art and some of them could have had personal experiences in art classes that were horrific. When teachers provide information and possibilities about their field to other teachers, they help bridge and reduce misunderstandings. One cannot assume that because one knows about and has a passion for one's subject area that others will too. Remember, other teachers may feel that way about their subject area as well, and it takes two or more individuals for relational practices to become enculturated.

Communication and the structures within which a team functions comes out of the strengths of the group members, agreed upon understandings, group values, and key functions. Shared vision sets the agenda and direction through consistent and constant team building including interactions that raise one another to higher levels of motivation. In order to build and nurture collaborative teams, we have developed home spaces within the workplace, established lunch clubs, book clubs, and knitting clubs, we work to share teachable moments, and to recognize individual achievements and successes. When members of a collaborative team share visions from personal values and strengths, they foster engagement and heighten accountability. Shared visions should then be displayed in the classrooms and planning spaces as visual reminders that serve as guidelines.

Here are a few criteria for team building:

Student Success. Promote learning, personal development, and academic performance through curricula, programs, and services that enhance student success.

Environment of Support. Provide supportive environments where members of the school community may freely learn, meet, live, teach, work, and relax.

Involvement. Foster a sense of community and engage students, faculty, staff, and community with the school through programs, services, events, clubs and organizations, employment, community outreach, and leadership opportunities.

Diversity. Promote understanding and mutual respect of personal and cultural differences and similarities by building supportive and inclusive communities and celebrating our uniqueness.

Advocacy. Assist students in navigating bureaucratic processes and systems and in reducing barriers to their success. Contribute to the community's understanding of students in and out of the classroom.

Strengthening Associates and Celebrating Victories

We mentioned resistance earlier but there are many kinds of resistance. One type

NOTES:

Advocating for Advocacy

> **Curriculum is a complex process that is in constant flux, negotiation, and development and it is not a product or a single learning episode for teachers and students.**

of resistance that has wreaked havoc in many schools in reformation is passive resistance. A prime example of a passive resistor is the case of Ms. Jones.[5] The teachers in Ms. Jones' small rural school decided to do an integrated unit. They contracted two artists to work with them and two professional development days were devoted to this project in which the artists and teachers planned the unit, goals, and timeline. The teachers were charged with introducing the concepts, skills, and historical information prior to the artists' residences. On the first day that the artists arrived, they met with each class. It was at that time that it was discovered that in Ms. Jones' class no work had been done. The students didn't even know about the artist residency. After that first day, Ms. Jones took sick days for the rest of the week. She had never questioned the project or process. She had attended the professional development sessions and her attitude always seemed agreeable. She made no complaints to other teachers. Her resistance was not apparent until her actions or inactions were already in progress, making this passive form of resistance extremely problematic to the collaboration. When such an incident occurs it is often best to regroup and explore why this is happening and create small steps to resolve the situation. In Ms. Jones' case, had the principal or fellow teacher spoken with her on that Monday and attempted to explore issues and options, the project could have gone more smoothly and might have served as a learning opportunity for Ms. Jones.

Frequent contact and cooperation among group members is critical in order to foster and sustain intergroup relations. Sufficient time must be provided for group members to overcome their initial feelings of anger and prejudice towards one another and develop trust. Relationships must be ongoing in order to effect long-term change. There must be administrative support to promote intergroup relations. Sincere investments of time and value of the process of collaboration develops stronger integrated curricula development, team climate, and student learning. This way of team building does not require more money, but it does require innovative, strategic planning. The example below is a five-year plan:

1. The first year we will invest in individual professional development that focuses on getting to know ourselves through workshops, book clubs, etc. The goal is for each teacher to develop one unit that includes a subject area not-of-their-own and invites the expert teacher into their classroom as a guest lecturer.

2. The second year we will invest in team development that focuses on getting to know the team through workshops, presentations, book clubs, and weekly planning time. The goal is for each team to create shared visions and values, to develop one lesson that includes other subject areas, and to invite that teacher into one's classroom as a guest lecturer.

3. The third year we will continue to invest in team development through shared professional development opportunities such as attending an external workshop or conference. The goal is for each team to review each other's standards and create an integrated curricular calendar. We will also continue to invite team members, when appropriate, into our classrooms, as stated above.

4. The fourth year we will continue to invest in team development through shared professional development such as attending an external workshop, class, or conference. The goal is for each team to create an integrated curriculum and continue inviting a guest lecturer into each other's classroom as well as possibly pilot some of the ideas

and lessons.

5. In the fifth year, we will pilot the integrated curricula and continue investing in team development through shared professional development. The goal is to revisit, reflect on, and revise the curricula.

Drake (1998) stated, "Creating innovative models of education values a different philosophy from that of the traditional, and it follows that all aspects of the educational environment must adapt" (pp. 189-190). She suggested some practical ideas that stimulate the celebration of visions and victories. The first is sharing stories—giving time to share the learning, the war stories, and the adventures, helps to continue the collaborative climate and provides bonding time. The second is for administrators to publicly recognize the team members for their accomplishments and to participate as one of the team's members when appropriate. The third component is to encourage the collection of data to document not only the process, but also the changes in students' behavior, attendance, scores, and increases/decreases in certain aspects of learning. This suggestion could also be turned into public relations possibilities for the community, professional presentations, and web-site information, as well as new funding and grant opportunities. For, it is not only telling each other our story, but learning to tell our stories that is important.

In conclusion, being a teacher in a school makes you a member of a team. Collaboration can come in many forms and depths, but collaboration is always the act of working together for a purpose. Collaborative teams help all members to find and develop their voices, leadership strengths, and similarities for a common vision. Strong collaborative teams also address the development of a sustainable, constructive, just, and equitable relationship among all group members. Effective collaborative teams engender relations and collaboration not only as an action for relations but also to serve as foundations for interdisciplinary/integrated curriculum. Investment in the development of a collaborative team is crucial to the foundation of integrated curricula. But the development of relationships and not curricula should be the main focus for collaborative teams especially in the beginning. That is, curriculum is a complex process that is in constant flux, negotiation, and development and it is not a product or a single learning episode for teachers and students. As Paulo Freire stated "Many times people assume I have the answers, but I do not—only the energy to explore" (as cited in Ballengee-Morris, 1998, p. 113).

... collaboration is always the act of working together for a purpose.

Suggested Discussion Questions and Activities

1. Read portions of Horton, M. & Freire, P. (1991). *We Make the Road by Walking: Conversations on Education and Social Change.* Philadelphia: Temple University. This book reflects the truest sense of association—a dialogue between two of the most prominent thinkers on social change in the twentieth century concerning their experiences, hopes, and dreams. Horton was the founder of the Highlander Folk School, a precursor of the 1950s and 60s U.S. Citizenship Schools. Freire established the Popular Culture Movement in Recife, Brazil's poorest region, and was active in educational development programs worldwide. How does this reading

Photograph by Donna Green, 2006.

affect your idea of collaboration?

2. It is your first year teaching and you have become a member of a team that has been established for some time. Often in the meetings, they refer to the former art educator saying that she or he never cared what was going on because she or he taught art and did not really educate students. How would you respond? What strategies would you develop to address these issues? Include administrative support, publications, ideas, and actions. Take a look at Lynn Sanders-Bustle's (2005) *NAEA Advisory* "Overcoming the Challenges of an Integrated Approach to Art Education."

3. Dr Mark Moilanen of Shippensburg University in Pennsylvania shared a unit plan created by Casie Geiswite, one of his art education students. **[Teacher contribution]** The unit entitled "Freedom from Fossils,"called for collaboration among high school anthropology, history, and art teachers in formulating activities that challenged students' to extend their study through creative explorations and visual creations of the prehistoric animals from which certain fossils and bones were derived. In this case, the art teacher looked to the standards and topics of the anthropology and history classes prior to instigating and ultimately leading this interdisciplinary curriculum project. From your research of standards of learning for a non-arts high school course, formulate an interdisciplinary plan proposal to share with other teachers in your school.

4. Below are questions that were mentioned in this chapter. Answer each question reflectively and as honestly as possible. The purpose is to reveal to oneself the whys, whats, and hows, as well as to identify the strengths that can be developed and offered to one's team.

a. Why are you teaching? What made you go into the field—a parent, a friend, a past teacher, always wanted to?
b. What motivates you to be a teacher? What stimulates you in the classroom? What is going on when you feel the students' excitement about the subject area?
c. What are your thoughts about learning? Do you consider yourself still learning?
d. Do you value mentoring? What has been a positive mentoring experience? What has been a negative experience?
e. Do you value learning about others?
f. How do you reflect those values in your daily tasks, decisions, and teaching?
g. What books or stories made the biggest impression on you as a child? What values did these books teach?
h. What books or stories are you reading right now? What values do they teach?
i. What strands of themes do you find in your answers that illustrate who you are as a teacher and how you practice those values?
j. What are your strengths? What do you have to offer to a team that goes beyond your subject area?

Once you have answered these questions, decide what you can share with others and make a separate list. Form a team and then compare and clarify answers and map out your team to find similarities and voids. This exercise helps to find who you are, what your strengths are, and how to make a team a functional, supporting, and sharing community.

Notes

[1] Ms. Walker is a fictional blend of teachers in author Christine Ballengee-Morris' former school.

[2] This is a fictitious story based on actual events in which author Christine Ballengee-Morris' was involved..

[3] Ms. Nana and Mr. George are fictional blends of teachers who have worked with Christine Ballengee-Morris.

[4] This magazine can be found on the Southern Poverty Law Center's website or by searching under Teaching Tolerance. It is free to anyone and the magazine's mission is to decrease all "isms." Lessons are available.

[5] This is a fictitious account of actual event witnessed by author Christine Ballengee-Morris.

Chapter 5

COLLABORATION AND STUDENT-INITIATED INQUIRY

Power. Teachers and administrators have it. High school students want it. It is as simple as that. We face this battle for power daily every time we enter the school. Some high school students reveal their fight for power through the clothes they wear or don't wear, the way they walk or don't walk, and to whom they talk and don't talk. And yet, high school is the center of the lives of many young people in the United States. High school students plan and get ready for school as they buy things, talk with their friends and family, write, read, research, work, and play. But, when they get to school, they often hate the experience. Many high school students detest rules. Most can't stand being told what to do and do not like—or find it extremely difficult—to sit still, be quiet, and listen. High school students generally want to be free to listen to music, be with their friends, and talk about love, life, and having fun. They hate being treated like children, especially when it seems that the teachers, administrators, and rules that govern the school dictate every aspect of their day which includes when they can talk, walk, sit, eat, and even go to the restroom.

In addition, many high school students may view learning and knowing as simply another direction imposed on them by the school and the system. Learning what the teacher tells them is only important as a means to get a good grade, pass the class, and eventually graduate and get out into the "real world."

The Art Room — First Days of School — Special Needs

Classroom Management — Instructional Strategies

Students in traditional high school art classrooms are typically faced with objectives that they have had no voice in creating. Although creativity and self-expression may be the focus of traditional high school art lessons, high school students need to develop the skills to understand and critically connect these learning experiences to realms outside of the art class. Employers and university professors lament the fact that high school graduates lack the critical, connective, and creative thinking abilities to solve problems, develop new ideas, and facilitate increased production, or profitability.

> As the growing unemployment of our most schooled workers demonstrates, academic success is at best irrelevant and may even be harmful to working productively in the real world. The exploding information base and intelligent tools of the modern economy make thinking skills far more important than the memorization of facts or rote exercises (Perelman, 1993, para. 15).

But, the idea of moving beyond traditional teacher and school directed approaches to teaching is not new. In 1938, John Dewey explained that all genuine education came about through connective experience. He believed that if students cannot apply what is presented in the classroom to their own lives and their other studies, then their power of judgment and capacity to act intelligently in new situations is severely limited. Dewey (1938/1963) wrote that education should not be merely a preparation for future life, but also should be full of life. His Laboratory School at the University of Chicago, established in 1896, involved students in practical aspects of life at the time among which included running a store, a post office, and a home. His work and writings were largely responsible for a drastic change in pedagogy that began in the U.S. early in the twentieth century as emphasis shifted from the institution to the student. Dewey (1938/1963) explained that every experience should do something to prepare a person for later experiences of a deeper and more expanded quality. He reminded teachers that, "connectedness in growth must be [their] constant watchword" (p. 75). He also challenged educators to arrange the kind of learning experiences for their students that engage rather than repel, ones that promote further inquiry and the desire to know more.

Teachers and Students as Co-learners

> A revolutionary leadership must accordingly practice co-intentional education. Teachers and students (leadership and people), co-intent on reality, are both subjects, not only in the task of unveiling that reality, and thereby coming to know it critically, but in the task of re-creating that knowledge (Paulo Freire, 1970/1994, p. 51).

Prior to the military coup of April 1, 1964, in Brazil, Paulo Freire (1973) acted as the coordinator of the Adult Education Project of the Movement of Popular Culture in Recife. His project, "a culture circle," replaced the teacher with what he called a coordinator. His pupils were referred to as group participants and the syllabi became known as learning units. This project analyzed aspects of Brazilian reality and was expanded into the field of adult literacy. The first attempt with a group of five participants was designed to be an introduction to democracy and culture. Critical consciousness replaced naive acceptance and participants were empowered to transform culture and intervene in history (p. 44). The group participants talked about multiple aspects, origins, and subjective mediation in the creation and recreation of culture. In the process, they began to understand that it was necessary to learn to read and write and become agents of their own learning before they could become critically active in changing their living conditions. Freire's work centered upon empowering the oppressed (impoverished South Americans) through a consciousness transformation by educating both oppressors and the oppressed for critical self-reflection.

Henry Giroux (1992) viewed education as a language of possibility in the form of human empowerment. Giroux advocated a radical education that refers to a kind of educational practice that is interdisciplinary, one that questions the fundamental categories of all disciplines, and has a public mission to make society more democratic. This complex approach to teaching is known as critical pedagogy. The theory that supports critical pedagogy views curriculum as a living practice that is centered on students' experiences (Brady, 1995). Jeanne Brady (1995) reminded us that a critical pedagogical approach to teaching could redefine the relationship between student and teacher by creating a classroom community of shared responsibility (p. 85). In other words, educators must see their classrooms as "not a neutral environment where knowledge is passed down from teacher to student, but as a complex social site, a borderland—a domain of crossing—that offers the possibility for mutual negotiation and translation where both teachers and students bring with them subject positions that are informed by their class, race, gender, and ethnicity" (p. 87). Through this process comes a transformation of the classroom as a site for celebrating the democracy of difference. This celebration encourages the voices of students to be heard—active voices, actively wondering and learning through a critical interrogation of knowledge.

Collaborative and Student-Initiated Interdisciplinary Learning

Students are liberated when teachers emphasize their lived experiences and share the responsibility of teaching and learning. At the same time, these approaches provide an excellent opportunity for interdisciplinary connections. The heterogeneous grouping of students in a typical high school art class is probably the most formidable obstacle to interdisciplinarity. Although basic requirements exist, students in high school often make important decisions about their future—often for the first time—as they decide which electives to take and begin advanced study in certain subjects.

The good news is that a teacher in a classroom environment that promotes co-learning, or embraces the idea of students and teachers learning with and from each other, does not have to make all the connections in isolation. The students will make the connections!

Students in a high school art class have a wide range of interests and avocations and rarely do all of the students in a given art class also take their other classes together. Even the most aware art teacher cannot know what is going on in all of the other classes her or his students are taking. The good news is that a teacher in a classroom environment that promotes co-learning, or embraces the idea of students and teachers learning

Photograph by Donna Green, 2006.

with and from each other, does not have to make all the connections in isolation. The students will make the connections. The bad news is that students are rarely asked explicitly to do so. A teacher who supports this approach will ask students to find interdisciplinary connections between what they are studying in art class and what they are studying in their other classes. Such connections can be simply stated and shared with the class in the form of discussions, critiques, written assignments or through images, signs or symbols in their artwork. When students are allowed and encouraged to make these kinds of interdisciplinary connections they greatly assist teachers in understanding what students are learning in their other classes.

Let us picture students in a typical high school art class for a moment.[1] A football linebacker straddles a stool and desperately tries to get his seemingly huge athletic shoes under the table. Across from him sits a young woman with jet-black hair, and safety pins up and down her shirt, and a few piercings in her eyebrows. She appears to be working very hard to look dazed and confused. A young man walks into the classroom, twisting and turning, while he holds up his oversized, baggy pants that seemingly could fall off at any minute. A grimacing boy follows him wearing jeans and a t-shirt and carrying only a baseball cap. A fresh-faced young woman with long red hair tied into a pony-tail, dressed in faded jeans, shirt, and vest sits at the table by the window. She is busy talking with another young woman

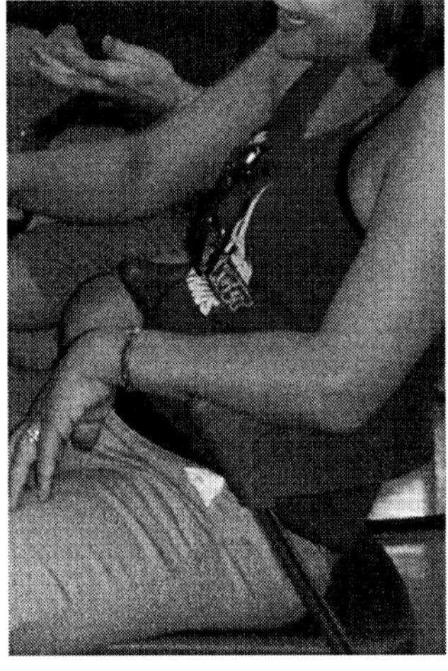
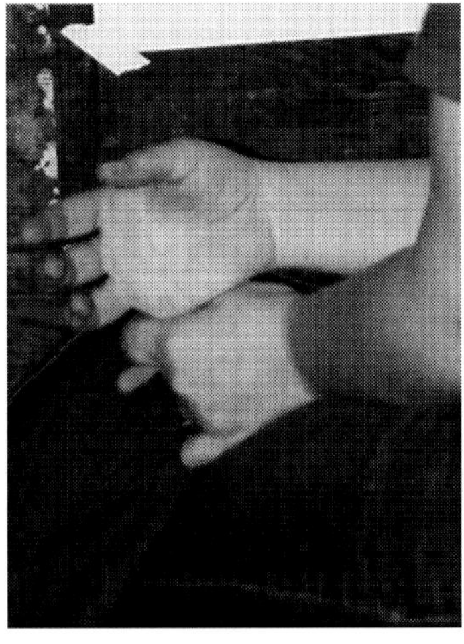
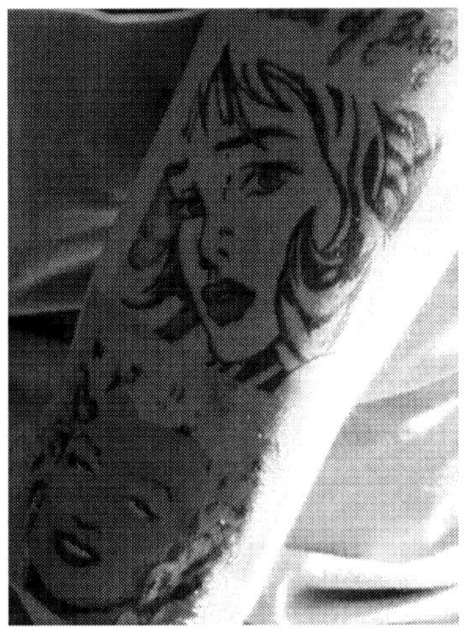

Typical high school students are as varied as the lives they live. But they are all there, in the same art class and each of them brings varied expectations, lived experiences, and expertise from which they all learn and teach each other.

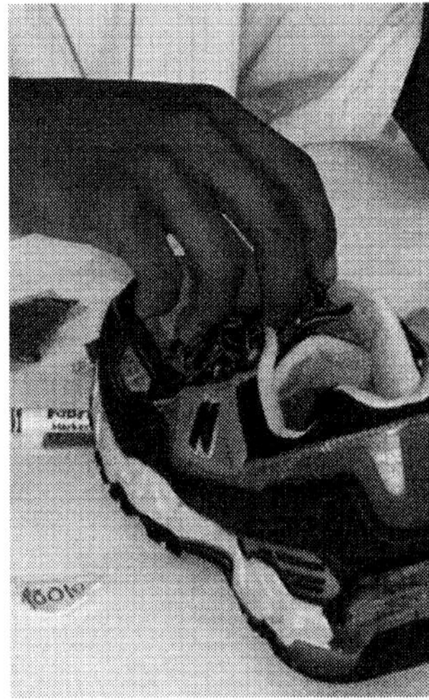

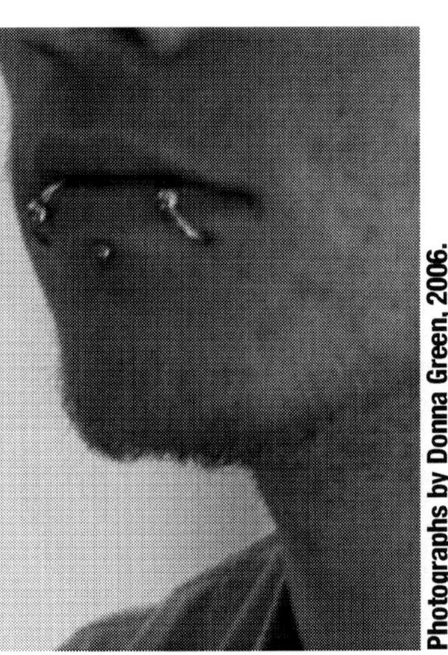

Photographs by Donna Green, 2006.

in spiked heels and mini-skirt who quickly adds more mascara to her eyelashes. A very small young man grins beside her while he is poked constantly in the ribs by another freshman boy wearing a professional football team jersey. There is also a young man in khaki pants and a white shirt with a pocket filled with pens and pencils. A cheerleader sits erect in her seat next to the window. Another young woman dressed in what appears to be her mother's suit, quietly watches the other students with her hands folded neatly in her lap the entire time. There are students with short hair, spiked hair, dyed hair, and many other styles. They are loud and quiet. Some giggle and others are somber. They are serious and silly, intense and flippant. Some are very involved in sports, some have jobs after school, and others are active in school and community clubs. They are budding artists, writers, scientists, actors, scholars, salespeople, teachers, athletes, chefs, farmers, police officers, soldiers, and artisans. Some have families. Some do not. Some have homes and others do not. Some care about life and living and some simply make it through the day. But they are all there, in the same art class and each of them brings varied expectations, lived experiences, and expertise from which they all learn and teach each other.

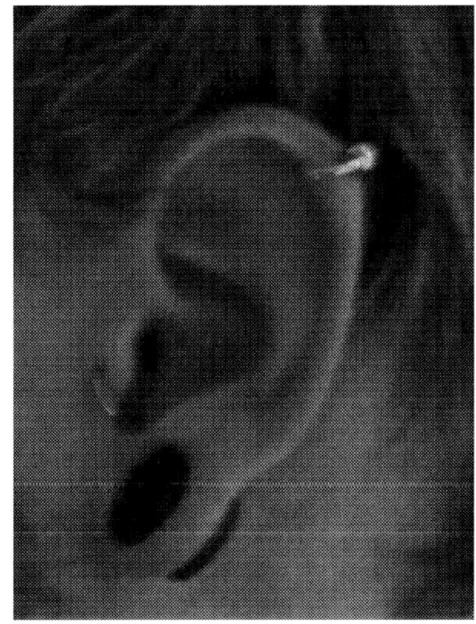

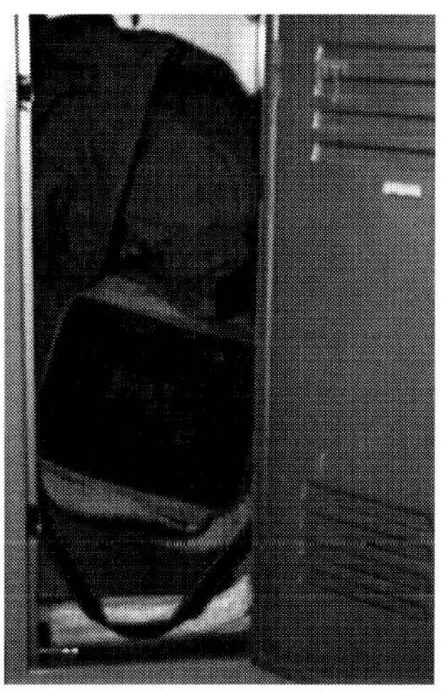

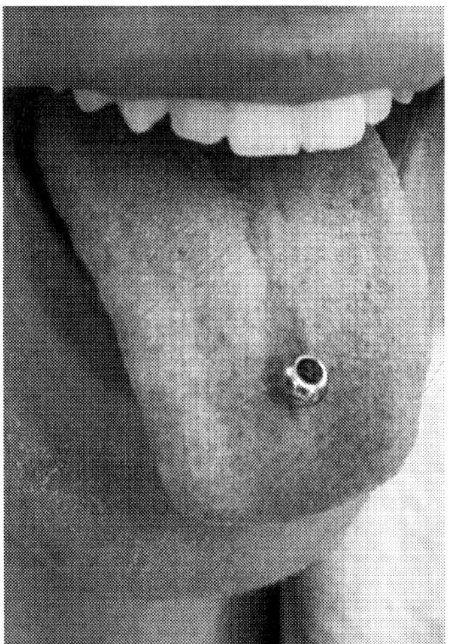

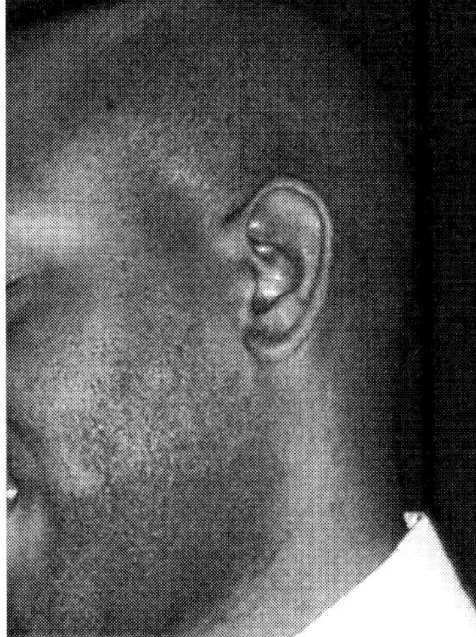

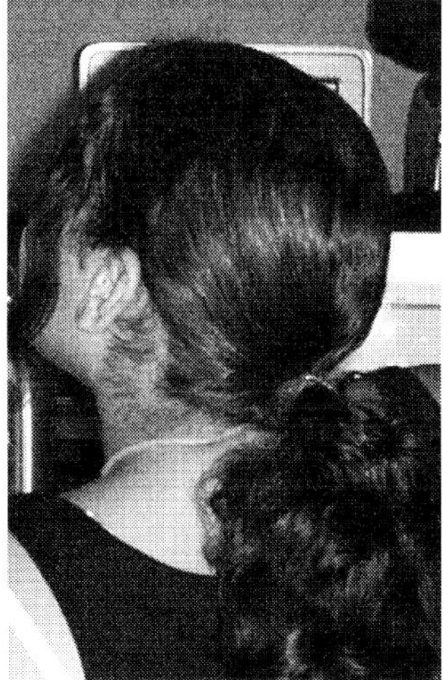

Students with Special Needs. Even in a typical classroom like the one we just described, student differences are not always easily recognizable. Some students may have learning challenges such as difficulty paying attention, problems with reading, writing, and/or simply staying in their seat. Others may have emotional, social, and physical challenges that affect how they learn and participate in art class. In many cases, students who are identified as gifted and talented are also considered to be special needs students. A number of supportive texts, workshops, and other resources are available for teachers to assist them in teaching students with special needs.[2] In addition, trained teachers and staff at all public schools, and in many private schools, are available to assist teachers who work with these students. In fact, this is their primary occupation. In our own teaching experiences we relied heavily on the assistance and expertise of the trained special education professionals in our respective schools. They provided us with individualized education plans (IEPs) for each student classified as a special needs student. Such plans instructed us to, for example, to create additional handouts with larger type for students with visual problems; give extra time on tests, or when doing any reading and writing activities for students who had problems with the written word; move a student to the front of the room or always stand close to a student who had a hearing or visual challenge; and allow a student with attention deficit hyperactivity disorder (ADHD) to get up out of their seat every fifteen minutes to retrieve materials and supplies. Special education teachers or aids are often assigned to accompany students with severe disabilities individually. When such an arrangement is not possible, teachers may assign a few students in the class to take turns and help a student who may, for example, have limited use of his or her hands or arms.

Students with special needs are a very important component of any class. Because of the individualized instruction often associated with art teaching and learning, counselors tend to place many more students with special needs in art classes than any other classes. The most important point to keep in mind in order for this situation to be most effective is that art teachers need to know if and when any of their students has a particular need that should be addressed or considered. Art teachers should take a proactive role to learn about the needs of the students their classes and should retrieve or seek appropriate documentation and direction from their administration. We recommend that all art teachers remain in close contact with the guidance department, administration, and special education departments in their schools regularly throughout the school year.

As we mentioned earlier, some students may be classified as having special needs because they are categorized as "gifted" or "talented."[3] Schools and teachers are charged with insuring that such students are challenged and inspired in all of their classes. Again, the individualized approach prevalent in much art instruction answers this directive in part. However, it is also important for teachers to pay attention and provide opportunities and possibilities for all students to go beyond the limits of the prescribed curriculum and planned learning activities.

Recognizing the value of each and every student in a high school art class is crucial in order to maintain an art classroom environment that is conducive to learning.

Classroom Management. According to Julia G. Thompson (1998) in her *Discipline Survival Kit for the Secondary Teacher,* typical high school students are intensely affected by peer pressure; have high emotional energy; believe having fun is the most important thing to do; want their education to have a practical purpose; do not want absolute freedom; are insecure; make mistakes due to inexperience; and want to be treated as worthwhile people (pp. 10-15). "The most powerful weapon available to secondary teachers who want to foster a favorable learning climate is a positive relationship with our students. When teachers interact in a collaborative and supportive way with students, pupils will respond in kind with fewer disruptions" (Thompson 1998, p. 6). Recognizing the value of each and every student in a high school art class is crucial in order to maintain an art classroom environment that is conducive to learning. With such value comes mutual respect, understanding, and attention to the ideals of the varied students in the class.

We recommend that such collaboration begins on the very first day of school. According to Ms. Donna Green, former art teacher at Mills E. Godwin High School in Henrico County, VA:

> When building your foundation, you may be able to draw from students' and other teachers' past classroom experiences. Ask students

Photograph by Donna Green, 2006.

to make suggestions about what should be expected of them and how misbehavior should be addressed. Students are often more responsive to rules they helped create. Creating an environment in which students know and follow the rules is challenging, but not impossible. With a little patience and perseverance, you can lay a foundation for respect and positive behavior in your classroom that lasts all year. (personal communication, February 6, 2006). [Teacher contribution.]

To achieve this goal, teachers may work with students to design the rules of behavior they believe are needed as well as the kinds of learning they expect in the class. The first day of a school year or semester may look something like the following example. First, the teacher introduces him or herself to the students and states the name of the class. During this introduction the teacher's name and the name of the class is clearly visible on the board, a poster, the television, or projected on the computer screen. Next, the teacher checks the class roster by asking the names of each of the students.[4] The teacher then asks the students what they expect to learn and do in this class. Some students may be very vocal in response to this request. Other students may sit quietly. As it is important for everyone's voice to be heard, the teacher may ask each student to contribute to the discussion by taking turns around the room or by asking the students to write their ideas and suggestions on a handout.[5] While the teacher should recognize the value of each student's ideas, he or she should also share with students that the objectives of each activity and unit must be intricately linked to the national, state, and school district's standards and curriculum. He or she may share some of those standards with the class and ask students to find connections between their own ideas and those standards. With experience, teachers can expand and/or change their pacing guides[6] or semester/yearly class plan to include the students' ideas and interests. Following the discussion of the class expectations, the teacher can ask the students what kind of classroom environment they need in order to accomplish the goals of the class. For example, if throughout the discussion, some students spoke loudly or there

NOTES:

The key to the success and student involvement in any learning activity, is interest and motivation.

was too much talk for anyone to be heard, the teacher or a student may point out this problem. The teacher will then suggest that certain rules will need to be established and ask the students to help formulate those rules.[7]

Pundits suggest that no more than five classroom rules should be developed (Canter & Canter, 1976). The teacher and students should discuss what the teacher needs in order to teach as well and what the students need in order to learn successfully. Cooperatively, the class and the teacher may develop rules such as:

1. Be in our seats and ready to work with all materials and homework when the tardy bell finishes its ring.

2. Listen and follow directions the first time they are given.

3. Keep hands, feet and objects to ourselves.

4. Use respectful language at all times.

5. Treat other people, art work, materials, supplies and equipment in the art room with respect.

Teachers who use this collaborative approach should address school and district codes of behavior during the discussion and guide the students appropriately in the design of their class rules. Notice that all of the rules are written as positive, rather than negative statements. Teachers sometimes find it difficult to encourage all students to be involved in the formulation of the rules on the first day, because many of the younger or quieter students might not join in the discussion. Teachers may call on students by name, ask their opinion of a formulated rule, or even ask them to judge whether a rule is broken in an invented scenario to help all students feel involved. We recommend that students and their parents sign copies of class rules, rewards, and consequences and that teachers keep these documents on file. This way, teachers have proof that students have read and are aware of the class rules and that they have agreed to abide by them. Some teachers include class rules on a perforated portion of the course syllabus. The form and syllabus become an advocacy tool for the art department because in addition to seeing the rules of the course, parents read about the exciting learning experiences that will happen in their son or daughter's art class.

As the school year continues and students become more comfortable in class, new teachers may find it difficult to simply call the class to order at the beginning, end or middle of class. We recommend that teachers and students agree upon a signal for this purpose rather than a less effective scenario, such as when a teacher attempts to raise her or his voice above the roar of thirty-five plus teenagers who are all talking at once. Some teachers find that switching the lights on and off, raising their hand, or holding up an object such as a paintbrush or hall pass are strategies that work well. Others report that giving a short quiz at the beginning of class (calling out "Question number one" as soon as the bell rings) assists them in establishing order. We have found that different teachers find their own effective ways to gain the attention of the entire class. The key issue to maintaining order and respect in the classroom is consistency. Teachers who command and expect order at the beginning of every class must also make sure that they give consequences when this result does not happen.

Another common complaint we hear includes students who quickly complete an assignment and refuse to work on further suggestions or directions given by the teacher. A well-designed unit of instruction includes objectives and criteria. As stated in Chapter 2, students should know and understand exactly what and why they are being asked to do what they are doing in class. Objectives and criteria for every exercise and work of art they create should be visible in the room. If and when a student decides that he or she is done, a teacher may simply refer to the list of objectives and criteria and ask the student to what degree they have met these requirements (or have a self-assessment rubric ready to mark).

In Chapter 10, we discuss the use of assessment models in which students are involved in the evaluation of their own work. Such models may also be used to inspire those students who appear less motivated to listen to teacher comments and directions. Of course, the key to success and student involvement in any learning activity is interest and motivation. Teachers must recognize, however, that even the best and most interesting lesson may not stimulate all high school students. Remember, there is always so much more going on in the lives of high school students outside of class than we know.

Rewarding good behavior is also significant to a productive class setting. Granted, while creating a safe and productive environment is a reward in itself, most of us like to be rewarded and recognized for doing a good job. We recommend that teachers create a reward system with the help of their students that takes into consideration the school and district rules and codes of behavior. For example, some schools do not allow food in classrooms. Therefore, a pizza party in the art room in this case would not be appropriate. One way a teacher could work around this issue would be to talk to the school principal and coordinate a pizza party in the school cafeteria. This sort of permission can be obtained prior to the first day of school. Or, if the students suggest a food reward, the teacher can say that they must first ask the administration before they can agree to this or any other reward.

Other rewards may include going outside to draw, listening to music in class, watching a movie, a free-drawing day, and/or a special activity such as an art auction. For example, one of the authors had a particularly difficult year with classroom management.[8] She decided to take an employer/employee relationship with her students and paid them for their work in class and fined them for breaking classroom rules. She created simple art money and used the copy machine to reproduce the currency. Each student received $5,000 a day and was fined $1,000 for each rule she or he broke.

At the end of the marking period the class held an art auction in the classroom, complete with a local auctioneer, food and drink, decorations, and music. Students had to follow the dress code (construction paper ties and hats were provided), pay $100,000 to attend, and needed additional art money to bid on the art they and their classmates had created over the semester. Students who did not have sufficient art money to attend negotiated feverishly with their classmates to borrow funds to participate in the event. This attempt to maintain order in a class became a huge yearly event that brought in administration, newspaper reporters, other teachers, and parents to the art classroom.

In addition to formulating rewards for good behavior, students should help establish consequences for when the classroom rules are broken. Such consequences could include a warning, detention, and/or recommendation for in-school suspension, as well as lunch detention in the art room. Parents or guardians may also be called. The consequences must also be aligned with school and district rules. Although additional cleaning duties in the art room during lunch time and after school may appear to be good consequences that can help the art teacher, we discourage this practice unless the students themselves have formulated this as a punishment. After all, cleaning up is an important aspect of art class and should not be considered a punishment. Teachers should remember that although they are sharing the formulation of classroom rules, rewards, and consequences with their students, they must reserve the right to deal directly with severe disruptions by sending a student immediately out of the classroom and/or calling in the authorities. Teachers should feel free to call on a principal or other teacher for assistance with minor or major classroom disruptions. Many schools and school districts offer discipline workshops for teachers or provide support for teachers to attend professional classroom management seminars. There are also a number of books and websites devoted to this most important aspect of teaching.[9]

Photograph by Donna Green, 2006

Although some materials and equipment must be kept under lock and key for safety reasons, we have found that basic art supplies that are kept in easily marked and accessible areas saves time and assists students in taking more responsibility for what they are doing.

The art room. The physical environment of an interdisciplinary, cooperative, and well-managed art classroom should be inviting and appealing. Student-initiated learning requires students and teachers to be actively engaged in the pursuit of knowledge while at the same time must remain cognizant of the rules and procedures they are expected to follow. Student ownership of the art room aids in the creation of such an environment. Students should hang their own and other's works of art on the walls. Interactive art displays where students both ask and answer questions daily make for dynamic classroom décor. Although some materials and equipment must be kept under lock and key for safety reasons, we have found that basic art supplies that are kept in easily marked and accessible areas saves time and assists students in taking more responsibility for what they are doing. For example, instead of asking the teacher for more paint or another piece of paper, students can just go and get what they need. We understand that some students can be very wasteful of supplies. However, when the students know and see that their class has a limited supply of materials to use on a given assignment, they may realize their limits and make appropriate choices and modifications. We know that some classes are very large and when thirty-five students or more get up and down from their seats the classroom environment can become very disruptive. We suggest that only one person from each table or group of five students be responsible for getting necessary supplies for her or his group. Likewise, when it is time to cleanup, we suggest that all students clean their own workspace. In addition, we recommend that, either by table or groups, students store their work and supplies and take brushes and other materials that need to be washed to the sink. Every class has different dynamics and what works in one class may not work in another. When teachers encourage student participation in what and how the class is being taught; formulation of the classroom rules, rewards, and consequences; and the way supplies are distributed, stored, and cleaned, they foster a highly productive and successful approach to managing their classrooms.[10]

One of the lessons teachers have learned from the reforms in education in recent years is that we will have much more success with our students if we focus on building positive relationships. The benefits of this are undeniable for teachers who are able to reach the kind of job satisfaction that they dreamed about when choosing a career and even more for the students who experience the thrill of an exciting school day where interesting discoveries are made and their opinions are valued. (Thompson, 1998, p. 8).

Student initiated connections. When allowed and encouraged to do so, students make and voice connections all the time. How many times have we heard a kindergartener shout out that tomorrow is someone's birthday? What was it about the class, the story, discussion, or activity that the student connected with this celebratory news? A Miro painting? Learning the days of the week? Similarly, a passing comment by a high school student during a class discussion, demonstration, or when toying with a paintbrush or paint container could inspire an exciting connective discussion. Granted, some connections and comments are more meaningful than others. The goal of student-initiated inquiry is to provoke the students to critically reflect on such connections, to look deeper and longer at the implications of the links they see between what they are studying or talking about in the art class and other realms of experience and courses of study.

In his book *Open to Question*, Walter Bateman (1990) explained the basic formula for inquiry teaching.

> Give students a set of facts and slip the leash. They speculate, they create new concepts, they apply old concepts, they test, they reject, they ask for more evidence. And the more of this process you can have students verbalize, the more conscious students will become of the way they are learning to infer, to test, to reject, to accept, and to help each other learn. (p. 18).

Ms. Jennifer Trettner is an art teacher at Ward Melville High School (Three Village Central School District) in East Setauket, New York. **[Teacher contribution.]** In collaboration, she and a colleague at a local elementary school involve their students in creating picture books.

> My students are given the challenge of writing, illustrating, designing, and printing their very own picture book. They are responsible for doing their own research on yearly topics. My students must use research, English, math, and technology skills, along with social studies and science knowledge (depending on the topic) in order to complete these books. Once finished, we visit the elementary school classroom for a day of sharing knowledge between the high school and elementary students. The high school students read their books to the elementary students, and the elementary students usually have their own presentation about the common topic to show us (personal communication, January 15, 2005).

Jodi Patterson, a high school art teacher in the Hemlock Public Schools of Hemlock, Michigan, allows an interdisciplinary approach to emerge within her classroom structure. **[Teacher contribution.]**

> Students are reminded there is no universal truth, no one story or no true aesthetic. This approach provides access to new experiences and transformational opportunities that are steeped

NOTES:

While an art teacher might extend the same assignment to all of his or her students, the works the students produce need not be identical, especially when those works function as representations and interpretations of ideas and responses that are important to the individual students.

Photograph by Donna Green, 2006

> in change within the self, student, teacher, society, and curriculum. A lesson may begin with a key student/issue or question for the class to depict in whatever manner they feel will best reach their audience, relay their message, and benefit as many as possible. (personal communication, January 22, 2005).

An example of Ms. Patterson's approach centers upon the question "What have we gained and what have we lost in the lifestyle we have adopted?" Her students talk about the question in circle-group-discussions. They complete a study-plan that requires a list of artists, theorists, or historians that they will research, the resources they plan to use, and a strategy for a creative or written product. Their final works of art typically cover a range of media, styles, and ways of working.

Ms. Patterson's approach brings to the forefront a very important issue in art class curricula—the teaching of media and technique. Although we recognize the importance of teaching and learning varied techniques of working in specific media, the clone-like "projects" that result from such a limited curriculum do not represent the "Art" we refer to in Chapter 2. In our mind, "Art" is thoughtful and challenging, and demands something of its viewer. That is, works of art are more than the material, formal, and visual qualities we see; works of art are about ideas, contexts, interpretations, and circumstances and to neglect these aspects is to deny our students rich opportunities to learn about the world. While an art teacher might extend the same assignment to all of his or her students, the works the students produce need not be identical, especially when those works function as representations and interpretations of ideas and responses that are important to the individual students. School art exhibitions should reflect the inventive and creative idiosyncrasies of student artists, not display work that all looks the same.

One of our most important tasks as educators is to help students enlarge the scope and depth to which they think about their lives. No matter where they live—in rural areas, suburbs, or cities—our stu-

> dents need our help to see that the world is a much larger place than the four walls of a classroom can ever contain. When we move our students into the larger world beyond their textbooks, we offer them benefits they can't learn by just sitting quietly at their desks (Thompson, 1998, p. 351).

Byrant[11] was the only African American student in an Art II class. He had recently moved from New York City to this small southern community and desperately wanted to talk about the differences he had found. Every time he voiced this to his class, the other students became defensive and angry. In Byrant's defense, it appeared that he was simply trying to understand. And in the other students' defense, they felt he was insulting them. As this constant battle became a threat to the learning that was going on in the class, the teacher began to formulate a unit of instruction surrounding a photograph with text by Adrian Piper, entitled *Political Self-Portrait #3 (Class).*[12] In this work, Piper typeset the story of her young realization of skin color and class and imposed the text over the entire surface of her photographed self-portrait. The story describes her visit to a school friend's home on 5th Avenue in New York City. After her visit, Piper rushed to her own Harlem home to tell her mother that she had found a wonderful place where her family should move. It was not until this point in her life that Piper realized that who and what she was dictated where and how she lived, as well as what she might become.

As Piper's work was displayed, Byrant and his classmates enthusiastically began talking about the various groups and/or classes of people in the school. They included geeks, African Americans, wannabes, Asians, do-gooders, preps, rednecks, artsy crowd, jocks, freaks, skaters, goths, cheerleaders, hispanics, European Americans, and dead heads. The students described the groups very carefully and when negative talk began, were allowed time to calmly defend and/or ask questions about specific groups.

> Some students are just not accustomed to speaking in a polite tone because they have not been taught at home the social skills they will need to succeed. Teachers should not condemn students for this failure of manners, but need to show them by example and gentle encouragement that respect must be mutual (Thompson, 1998, p. 15).

Some students were surprised that what they thought about a particular group was not necessarily true and how stereotyping, judging, and labeling people through mere observation or hearsay is not a correct or fair analysis. Many of the students shared this discussion with the social studies teacher and continued this line of exploration when learning about and discussing cultures from other countries. A few of the students worked with the social studies teacher to create a special assembly in celebration of Dr. Martin Luther King, Jr.

The Collaborative Art Teacher

Teachers who welcome learning whether it be from a professor, colleague, student, or child, are exciting people to be around.

> There is usually at least one in every school. Even in the most violent and out-of-control schools, there is usually at least one classroom that is a haven of peace and productive behavior. In this room, students engage in interesting learning activities that keep them coming back for more (Thompson, 1998, p. 29).

Students respect teachers who receive an obvious joy in what they do. Such teachers make sure everyone is welcome. They are organized but not afraid of change and they readily admit when they are wrong or when they do not know an answer, reason, or solution. A good collaborative and interdisciplinary teacher seeks answers and solutions, while at the same time is open to other's connective and inventive ideas. Those ideas may come from other teachers, students, administrators, people on television, or cafeteria and custodial staff. The persistent tone of mutual respect from a collaborative, interdisciplinary art teacher provokes others to want to be involved in the exciting things going on in the art classroom.

Instructional Strategies. Exciting and varied instructional strategies are part of the good collaborative and interdisciplinary art teacher's reper-

Photograph by Donna Green, 2006

A high school student's interest may be sparked or a disciplinary problem may be prevented when a teacher imposes a simple change of perspective.

toire. Unlike most elementary and middle school aged children, the majority of students in high school are bored easily with routine. They need change, alternatives, and differing approaches to study in order to maintain focus.

Direct and indirect teaching are two main instructional models. Best suited to teaching facts and standard procedures, direct teaching involves lecture; demonstration; daily review and checking; presenting and structuring new content; guided student practice; feedback and corrections; independent practice; and regular reviews. Indirect teaching approaches the learning process as inquiry. Students develop thinking skills as a result of indirect teaching that encourages student inquiry through discovery. In such an environment, students and teachers consider the learning context as a problem that they must address.

In her book *Design for Inquiry: Instructional Theory, Research and Practice in Art Education,* Elizabeth Delacruz (1997) described a number of instructional strategies that serve to vary and engage direct and indirect teaching in the art class. **Questioning strategies**, according to Delacruz should be clear and purposeful. She suggested that teachers ask questions of the entire class first, and then give time for everyone to formulate an answer before calling on specific students. Teacher feedback to students' answers should motivate further interest rather than simply acknowledge an answer as right or wrong. For example, instead of saying "Yes, that is correct" or "No, that is incorrect," teachers should ask another student to elaborate on this answer. Teachers might also pose another question that may relate something relevant in the student's lives such as "What television shows have you seen that deal with similar issues of identity as those we discovered in Adrian Piper's art work?" Another option would be for teachers to restate student answers in a form of a question such as "From my understanding, you are saying that a self-portrait does not have to be a picture of the artist. How would Adrian Piper's meaning change if she pictured only the coat?" We must also remember that in many cases, there may be no absolute "correct" answer. We want our students to feel comfortable and empowered to ponder new ideas and ways of thinking while they formulate and share these ideas with others. "Teachers need to respond to all students with specific, useful feedback, even when student responses do not necessarily coincide with teacher's 'questioning plans' or mental scripts of how the discussion should be going" (Delacruz, 1997, p. 45). For example, Byrant's need to discuss regional differences in racial attitudes inspired not only a lesson change, but also enlarged the new lesson to encompass geographical influences on culture and history. Bryant was able to share and correlate his own recognition of difference as a result of moving to Virginia from New York with Piper's story. His classmates also correlated their recognitions of difference from their own experiences to Piper's work and Bryant's contributions.

Delacruz and others (Adler, 1983; Davidson, 1952; Elkins, 2000) referred to **Socratic Dialogue** as an instructional strategy in which students must reflect and think independently and critically. Based on Plato's *Protagoras*, in which Socrates' re-examines learning as he calls into question the worthiness of seeking knowledge based solely on everyday popular thinking, Socratic Dialogue involves "a critical re-examination of one's accepted habits of thought and conduct" (Davidson, 1952, p. 19). Through Socratic Dialogue, students penetrate the world of appearances—what they like and do not like, what they think and believe—to question themselves as they look deeply and critically. In the process, students can develop questioning, listening, and seeing skills that require them to "hear the confusion in their own voices and see how they 'misread' texts" (Elkins, 2000, para. 38). "Philosophy begins when one learns to doubt—particularly to doubt one's cherished beliefs, one's dogmas and one's axioms. There is no real philosophy until the mind turns round and examines itself" (Durant, 1962, p. 9). For example, Ms. Patterson's students were very involved in reflecting on their own beliefs as they researched and responded to the challenge of thinking beyond their perceptions of universal truths or accepted dogma.

According to Delacruz (1997), **Cooperative Learning** is an indirect strategy to foster group inquiry. Cooperative learning is different from mere group learning in that students are heterogeneously grouped. In this approach, rather than work on one project, the goal of the group relies on learning by all members and all students learn the same material, which is usually provided by the teacher. In a cooperative learning strategy there are no group leaders. "Cooperative learning incorporates both direct and indirect teaching strategies. It involves the transmission of information from teachers to students as well as the communication of ideas between students" (Delacruz, 1997, p. 47).

In Bradford, Vermont, Oxbow High School art teacher Mary Chin "team teaches" an interdisciplinary art and music course with the music teacher. **[Teacher contribution.]** If, for example, the teachers and the students work together to search for and create a symbolic woven fabric that represents the many varied ideas and cultures of the peoples of the United States, they are involved in cooperative learning.

NOTES:

Nicolas Roukes (1988) described **synectics** as an instructional strategy that involves creative thinking, imagination, and analogical thinking. Similar to free and associative thinking, synectics embraces thoughtful connections that make meaning and/or alter original objects, intentions, and impressions. When art teacher Kathleen F. Ragusea worked with the Delta Program in State College, Pennsylvania, she taught an interdisciplinary course entitled "Art and War." [Teacher contribution.] As she explained, "My students examined the relationship between art and power in case studies that unfolded across history, for example: art of ancient societies; Goya vs. David in the Napoleonic Era; the revolt of the avant garde during WWI; and the totalitarian subversion of art in Hitler's Germany" (personal communication, January 21, 2005). This lesson would support synectic learning if some of Ms. Ragusea's students looked at the social implications of such monumental art as the Parthenon when they connect art and power (for example, could the building have been completed if the workers had not been enslaved?). Synectics involves meaningful yet divergent connective thinking.

A less formal, but nonetheless very effective instructional strategy involves simply **changing the view or perspective of the teacher and the learner.** As actor Robin Williams demonstrated in the film *Dead Poets Society* when he and his students stood on their desks, a simple change in a teacher's or student's physical location or stance can influence their ways of perceiving and knowing. A high school student's interest can be sparked or a disciplinary problem can be prevented when a teacher imposes a simple change of perspective. Sometimes teachers can impose just enough change if they require the class to sit on mats under the tables, remove tables or chairs in the room, move to a different location inside or outside of the school, and yes, even stand on a desk or counter (when appropriate).

Change for the sake of change may be too disruptive, however it can be an effective instructional technique when such a change in view or perspective relates to the interdisciplinary nature of the lesson. For example, in the perspective lesson described in Chapter One, on one day the teacher[13] held class under the tables and projected images of Raphael's *School of Athens* and Faith Ringgold's *Sunflower Quilting Bee at Arles* on a large paper attached to the edge of the table. The next day, the class was held in the highest point of the school, the loft of the second floor library. These different locations refer to the vantage points depicted in these two works, and students discussed the way their immediate perspective was changed in these two experiences as a result of their eye-level, physical location, and comfort level.

Another important instructional strategy in this unit on perspective involves **student inquiry and research**. Students generated questions surrounding a particular work of art or idea and then found the answers through deliberate and intensive research activities. While viewing Raphael's *School of Athens* and Ringgold's *Sunflower Quilt at Arles*, students asked "Who is Sojourner Truth?", "Why are Plato and Aristotle at the center of the *School of Athens*?" and "Why is there more information in the encyclopedias on the figures in Raphael's painting and little if any on the women in Faith Ringgold's quilt painting?" In response to their own questions, the students conducted research that took them to the library, the World Wide Web, the community, another teacher's classroom, and an administrator's office.

When teachers divide class time with **In-Process Critique** questions that tie in with big ideas and objectives of the unit, they assist students to take time for critical reflection about what they are doing in their art making activities. Similarly, mid-class time can be used to re-emphasize, clarify objectives, or discuss a potential change or new opportunity. During an in-process critique of the perspective unit discussed in Chapter One, students shared who they selected to represent in their artworks as the important figures in their lives and why. Some students may take this opportunity to rethink their choices, change them, and/or ask for help and assistance in figure drawing technique. For example, in the unit of instruction based on identity and artist Adrian Piper, the class decided to abandon their original plan of photography and create colored pencil drawings over their stories instead.

Figure 5.1. Joseph Norman, *Dangerous Gardens* (triptych), 1999. Lithograph, sheet size: 42 x 30 in. each panel. Museum of Art, Rhode Island School of Design. Permission of the artist.

Art as School Resource. Many high school art teachers are approached daily with requests for "art projects" that will fulfill the needs of the school. From signs to programs and brochures, such requests can often become opportunities for interdisciplinary discovery. For example, creating a sign for a school snack bar could require students to work with the geometry teacher to better understand grid enlargement proportions. When designing programs, posters, sets, and costumes for school concerts and plays, art students and teachers work closely with music and drama students and teachers. Art students may read plays and watch rehearsals to help them formulate ideas for images while music and drama students may come to the art class to look over sketches and designs. When working with robotics, physics students may request help from the art class to design "skin" or outer aesthetic features of their robots. The work of both the physics and art classes could be connected and informed by looking at and studying common works of art by such contemporary artists such as Elizabeth King[14] who employs the use of robotic designs in her artwork. Requests for posters or art work for classrooms and offices throughout the school building could involve art students in creat-

The single most important tactic teachers can use to approach the high-spirited, independent, and challenging adolescents who make up our high school art classrooms is to provide an atmosphere that provokes them to take an active role in their own learning.

ing signage or text panels that provoke links between what is going on in a particular space and the work of art they choose to hang there. For example, a poster of contemporary artist Joseph Norman's[15] *Dangerous Gardens* (See Figure 5.1) series or a work of art by a student that was inspired by this work could be displayed in a science classroom, and the science teacher could post questions related to ecology next to the work. Science and art students could also work together to create ephemeral outdoor sculptures (Taylor, 1997).

Often, the requests the art teacher receives from other teachers involve materials and supplies. In some schools, art classrooms are seen as storage sites for everything from construction paper to markers. Although we recognize the need to be protective of art supplies, we suggest that art teachers become interdisciplinary-minded when such requests are made. Why does this teacher need construction paper? How can the activity he or she is planning to do provoke a learning opportunity for art students as well?

Art teacher Kathleen M. Thompson (1995) explained that as non-art teachers lack the time and expertise to suggest art connections, it is important for art teachers to take an active role in instigating interdisciplinary collaborations. Students and teachers can discover exciting, connective opportunities through simple discussions, visits to other classrooms, taking note of what other teachers place on the bulletin boards outside their classrooms, or inviting other teachers to come to the art room. As Ms. Thompson (1995) reported "art enjoys a higher status with fellow teachers in an interdisciplinary setting" (p. 45).

Conclusion

Teaching high school students is very challenging and yet, from our own experiences, it is one of the most rewarding vocations we have experienced. We agree that probably the single most important tactic teachers can use to approach the high-spirited, independent, and challenging adolescents who make up our high school art classrooms is to provide an atmosphere that provokes them to take an active role in their own learning.

> The best teaching methods and strategies are those contributing to a climate that fosters self-confidence and encourages critical self-inquiry and self-reliance. The real question is not whether students appear to be busy on tasks or free to discover and invent, but whether they feel safe to take risks, whether they are willing to engage unfamiliar conceptions and connect those ideas to what they already know, and whether they are provided necessary conditions to learn how to learn (Delacruz, 1997, p. 50).

Many factors contribute to the creation of an empowering and inspiring classroom atmosphere. When teachers use varied instructional strategies they help prevent lethargy and boredom in their students. Teachers who share ownership of the classroom, supplies, and equipment promote responsibility and community. The management of a classroom atmosphere where students recognize the value of a code of conduct as well as the rewards and consequences of behavior is essential to a productive, interdisciplinary learning environment. High school art teachers who actively listen to and share with other teachers, students, administrators, and staff in the school as well as in the community, serve as role models for seeing and embracing connective ways of knowing. Most importantly, when teachers earnestly embrace student ideas, listen actively, and honestly take into account the values and interests of their students, they encourage meaningful avenues for interdisciplinary connections.

Suggested Discussion Questions and Activities

1. Students in a typical high school art class range in age, interest, and maturity level. What teaching strategies or ways of working can you think of that would individualize study while also interest, challenge, and engage students in interdisciplinary thinking?

2. Classroom management skills are essential to teaching any subject. Conduct further research in this area and explain what you would do in the following scenarios:

The tardy bell has sounded and your class is still very loud and disorderly.

Jimmy refuses to work on his still life drawing. He says that he is not an artist, can't draw and doesn't want to do it.

Mary and Joe had a fight this morning before school. They continue the process when they come into your class, disrupting the class so much that you cannot teach.

Kelly pours paint all over the table and laughs.

John is tardy yet again to your class--the fifth time this semester.

Richie and Tommy make derogatory remarks about Joe, his clothes, and his family. When you say something to them, they say they are just kidding.

Yolanda is lethargic, not responsive to you or the lesson, and seems to get aggressive every time you try to include her.

David tells you that another student in your class has a weapon.

3. Create a poster with your classroom rules, rewards, and consequences for your teaching portfolio.

4. Incorporate at least three different instructional strategies in an interdisciplinary unit or lesson plan that you have created.

5. Read a unit of instruction written by another teacher. Describe at least one way a student might change or initiate a different interdisciplinary learning direction in the unit.

6. Research classroom design and create one that would inspire interdisciplinary thinking for your future high school art class.

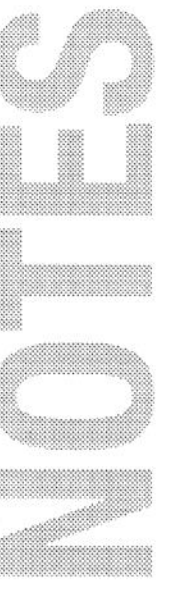

Notes

[1] This description is based on all the authors' experiences both teaching in high school and visiting high schools over the past 20 years.

[2] For more information regarding teaching students with special needs in the art classroom see Nyman & Jenkins, 1999.

[3] For more information on teaching "gifted and talented" students in art, see Clark and Zimmerman, 2004.

[4] Learning student names becomes easier with practice. Assigning seats and then creating a diagram with student names is a good way to both remember names and take attendance. Some teachers play name games and ask each student to share something about themselves like a favorite movie, restaurant, or musical group. Making an association between a student's name and their appearance, someone they look like to you, or with a song or story also helps.

[5] Using a first day handout with questions about what they expect the class to be, what they want to learn about as well as some philosophical questions such as "what is art?' make for important ways to get to know your students. Such information may also contribute to future teacher action research.

[6] A pacing guide is a semester or yearly plan that outlines all units and lessons that will be taught, resources, time involved, and standards/objectives that will be taught.

[7] Some high school art teachers reported that they preferred to address rules of behavior on the third or fourth day of class to avoid their rules being confused with their students' other classes. What is important to remember is that although each teacher develops her or his own way of managing a classroom according to her or his teaching style and personality, it is imperative to listen and give voice to high school students in this process.

[8] This story is based on author Pamela G. Taylor's first year of teaching art at Christiansburg VA High School in 1988.

[9] For more information regarding classroom management see Johnson 1998, Susi, 1995; Canter & Canter 1976; and http://users.pandora.be/education 2004. Also, a simple search in bookstores, libraries, or on the Internet will result in numerous sources of classroom management information.

[10] For more information regarding art classroom design see Goodwin, 1993.

[11] Byrant was a student in author Pamela G. Taylor's high school art class in 1996.

[12] A reproduction of this artwork can be found in *Contemporary American Women Artists* published by Cedco Writing Group, San Rafael, CA. For a complete lesson plan, see Taylor, 2002c.

[13] Author Pamela G. Taylor was this art teacher.

[14] More information about artist Elizabeth King can be found at http://www.kentgallery.com/kin.htm

[15] More information about artist Joseph Norman can be found in Taylor, P. G. (2002a). *Amazing Grace: The Lithographs of Joseph Norman*. Boston: The Center for Afro-American Artists.

Chapter 6

A CALL TO "FACE UP" TO CULTURAL DIVERSITY

One day as Mr. Blankenship[1] and his students returned from the parking lot after being interrupted by a fire drill, he saw two students standing in the hallway outside of the art classroom. One student, Mike held a driver license and laughed while the other student, Hope seemed confused and surprised. In fact, Hope folded her arms as if she were insulted. As Mr. Blankenship approached the door he heard Mike say, "This picture makes you look white but you're really black." Hope responded, "Well, I am both, sort of. My mom is white and American Indian and my dad is black but he is white, Indian, and other stuff, too. You can't show everything I am in one picture." Mike returned the photograph to Hope and they both entered the art classroom and took their seats. Their conversation inspired Mr. Blankenship to change his plans for the remainder of the class period. He opened an Internet browser, typed a few words in the search engine, and found an image of a black and white photograph by James Luna. (See Figure 6.1). The image depicted the artist facing forward with long black hair and no moustache on the left side of his head and a moustache and short hair on the right (First Nations Art, 1999).[2] Mr. Blankenship remained silent as the students found their seats, examined the image, and began talking. "He looks funny!" one student shouted. "Why did he do that?" asked another. "He's Mexican and Indian at the same time, duh," announced a third. Mr. Blankenship continued to sit, look, and listen, impressed by the increasing sophistication of the questions his students asked about the artist, his motivation for creating the work, and about the meaning of the work. Mr.

Multicultural Pedagogies — Aesthetics and Perspective

Cultural Diversity — Cultural Identity Formations

Blankenship spent the rest of the class period in conversation with Hope, Mike, and their classmates about the cultural and racial diversity of their community, what it was like to live as a mixed-race American, and the various ways the students in the school shared "who they are" with others though their clothes, hairstyles, and music.

So far this book has been about interdisciplinary/integrated curricula and yet this chapter issues a challenge to face-up to cultural diversity. "What does diversity have to do with integrated curricula?" Our answer is complex and yet simple. **Diversity is about everything**—how we teach, what we teach, who we teach, who we are, the communities we live in, our past, our present, our future, and our students' past, present and future. It is about our lives, the world we live in, and the world makers we teach.

As we have said before, the focus of this book is to encourage and indeed to challenge high school art teachers to take an interdisciplinary/integrated approach to teaching and learning in and through visual art. This process involves a transformational tactic that challenges us to think about not only what we do, but why we do it—not only what we teach, but why we choose to teach it. Our choices are determinate and influential. They give voice to some and silence others. Our choices reveal while at the same time they promote bias. Put another way, as educators, we need to critically and reflectively shift the paradigms that inform and promote multicultural education. That is, we believe that teachers who approach interdisciplinary/integrated ways of teaching need to question "what is not there" and "what is not done" as they rethink the pedagogies, content, and outcomes of their curricula. The solution is not as simple as tacking on "special units" about American Indians or African Americans or inserting a theme or two about shelter or rituals of other cultures from some ancient civilization. When these responses are enacted, the diversity has not been accomplished. That is, these responses are just other forms of marginalization and racism. In this chapter we explore what diversity is, and examine multicultural pedagogies, aesthetics, and cultural identity formations that make diversity content inherently interdisciplinary. Before going further we admit that facing up to the vast yet essential implications of cultural diversity in art and education is beyond the scope of this chapter and indeed, it is beyond the limits of any one book. The very nature of cultural diversity is its multiplicity. Therefore, in an attempt to suggest strongly that our readers use the information in this chapter as an impetus for further study, we provide parenthetical resource links both as citations and as explorative recommendations.

Diversity intersects all subject areas. An individual's varied experiences within the history, heritage, tradition, and culture of the groups to which they belong produces diversity (Ballengee-Morris & Stuhr, 2001; Chanda & Daniel, 2000). Paulo Freire (1998a) in his book *The Pedagogy of Freedom* recognized the interplay of diversity and education and made explicit the importance for teachers to model respect and the inclusion of many. This chapter is an exploration of multiple perspectives of cultural identities through interdisciplinary/integrated curricula. The purpose of this chapter is to establish a lifetime dialogue about cultural identities, differences, similarities, and social responsibilities and to encourage students and teachers to inquire how and why these components are visually presented and consumed by society.

Diversity—What is it?

Part of any discussion of diversity issues must be grounded in the concept of multiculturalism. Since the turn of the 19th century, people have interpreted multiculturalism through various lenses (e.g. sociocultural, political, academic, and pedagogical), all of which are biased (Daniel, Stuhr & Ballengee-Morris, n.d.). As reality continually shifts, philosophers, scholars, cultural critics, and social scientists question the evolution of multicultural. The breadth of the conceptual changes that surround the term multicultural is profoundly affected by and, in turn affects visual culture and education as represented through the arts and contemporary media and artifacts.

The complex issues of cultural diversity are studied often as a part of the school reform movement known as Multicultural Education that originated in the 1960s as part of the Civil Rights Movement to combat racism in the United States. It was then, and still is, an educational process dedicated to providing more equitable opportunities for disenfranchised individuals and groups to gain status and voice in social, political, and especially educational arenas (Sleeter & Grant, 1988). All forms of education act as social intervention and the implementation of these forms reconstructs society in various ways.

Paulo Freire (1998b) in his book, *Teachers as Cultural Workers: Letters to Those who Dare Teach*, wrote an eighth letter that explored the relationship of cultural identity and education:

> The identity of the subjects has to do with the fundamental issues of the curriculum, as much with what is hidden as what is explicit and, obviously, with questions of teaching and learning (p. 69).

Concepts of multiculturalism in art education should encourage social justice and thriving, constructive communities. "This type of education is a process guided by democratic social goals and values that also helps explain and confront colonialist practices, which stem from one group of people having power over another group's education, language(s), culture(s), lands and economy" (Ballengee-Morris, 2000a, p. 102). Moreover, these educational goals and values provide opportunities for classroom practices that build empathy, democracy, and social justice. But how do teachers design interdisciplinary curricula that work toward these goals? One way to begin is to consider the following questions: Do you

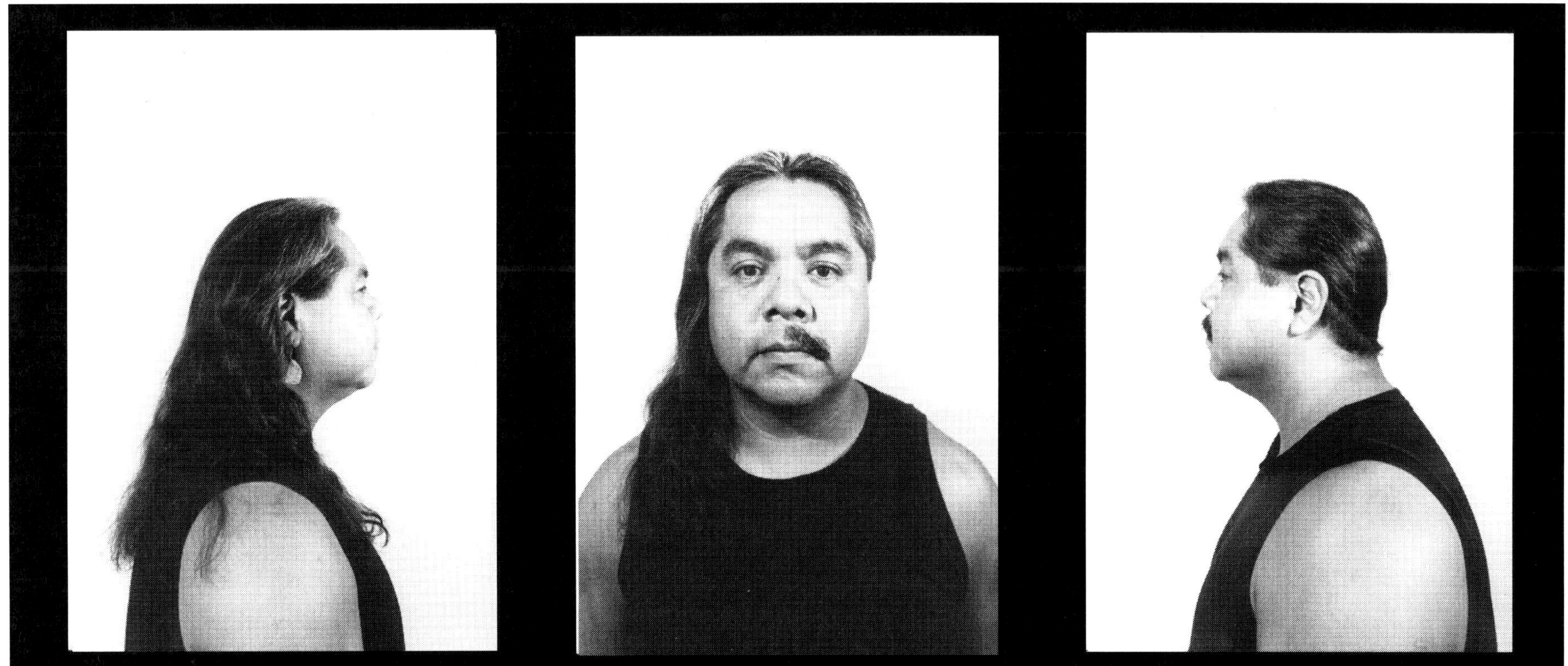

Figure 6.1. James Luna, *Half Indian/Half Mexican,* 1990. Permission of the artist.

as a teacher use a variety of images and diverse artists that represents a wide range of multiculturalism? Do you tend to group images of works by "minority" artists into celebratory weeks—American Indian week, African American month and so on—without ever viewing the images or talking about the artists in a broader way as one might talk about "dominant artists?" One way to counter this type of multicultural educational practice would be to present a variety of artists and their works, and to question oneself about when it might or might not be appropriate to speak about their diversity.

Good interdisciplinary/integrated curricula and teaching, particularly in the area of multiculturalism, should connect with the students' needs, experiences, and communities (Daniel, Stuhr, Ballengee-Morris, n.d.). It is important for students to investigate how their lives connect to and are limited by the broader society.

Students need critical skills to address social issues and to think through how some groups may benefit or suffer as a result of colonial practices and the decisions of other people. As an example, let us take an essential question: "How do stories shape access to privileges?" A teacher in Ohio[3] explored this question with her high school students in a unit of study about the social, cultural, and political issues surrounding the Octagon Mound in Newark, Ohio. (See Figure 6.2) In this situation, the two-thousand-year-old mounds constructed by American Indians are now owned by a historical society that leases the property to a private country club. For several years, American Indians have protested for access rights to the Octagon Mound, but the press and local and regional citizens seem not to have understood the complexities of the issues (Ballengee-Morris, 2006). Through surveys, the high school students found that most citizens had derived their opinion about the situation from what they had learned in the Ohio history books. Published for both fourth and eighth grade levels, the books state that the Native people who built the mounds no longer existed and the name given to those people were Hopewell and Adena. Most citizens did not understand why the contemporary

Figure 6.2. Aerial photo by Richard Pirko, 1994 (Youngstown State University). Permission of Pirko.

Natives were so upset over a situation that was not related to them—why should a Shawnee or Cherokee person care and what does that have to do with them now? The students proceeded to interview scholars, Native people, and informed citizens.

They learned that due to colonial practices, the historical society named the mound makers not by their known names, but by the discoverer of the mounds or relics of the area. And when the identity of the makers was investigated in the 1700s, a few American Indians were asked and no answers were given. The students found that there were no Hopewell or Adena Indians in the area and that no one had ever thought of asking Native people what their oral histories told of the mounds. From this small exploration, the students learned about the importance of voice, lack of voice, historical significance when viewed as the truth, the power of the press, and privilege. This exploration also helped the students understand better the complacencies of the local community about this and similar issues related to American Indians.

Figure 6.3. Author Christine Ballengee-Morris' contemporary Native regalia, 2006.

Like the Shawnee and Cherokee, many indigenous groups today suffer from archaic definitions of their cultures and art forms. Much of this inaccurate stereotyping is due to the way these cultures are taught in the schools. The market is full of lesson books that give simple cure-all examples of activities for teachers to use in the classroom. We do not advocate teaching to or about groups of people in only a celebratory way, because such an approach can be harmful to many peoples for a variety of reasons. No one represents everyone in a cultural group. There is no one thought or one way all people in a cultural group think, act, or respond, and there are multiple ways of knowing and doing within cultures (Ballengee-Morris, 2006). (See Figure 6.3)

Identity Development

Cultures are not discrete entities but are connected to peoples' lives and driven by their choices. People within groups, who are part of a state or nation and who are influenced by global events and media, carry the responsibility of their rights and can be blamed for inaction. We believe that teachers, students, and the larger community must think critically about their own and their group's actions—who they empower or disenfranchise through their personal life, actions, and work, which includes making and interpreting the meaning of visual culture (Ballengee-Morris &

Stuhr, 2001) because the action or inaction of each person ultimately affects all persons. **[Chapter 7 is devoted to service-learning.]**

Culture is a multifaceted and complicated concept. Many people believe culture to be a static, esoteric entity that is outside of an individual's lived experience. According to Daniel (2001), culture is made up of what we do and what we value. David Morris, an Appalachian cultural Ohio artist-in-residence, said, "[C]ulture is the heritage of the future" (Morris, in personal communication, 2000). Culture provides a dynamic blueprint for how we live our lives and confines our possibilities for understanding and action. We all have culture because we all live and exist within social groups. How we live our lives is influenced by aspects of our personal sociocultural identity as lived within a particular nation or nations and influenced by global issues.

The aspects of one's personal cultural identity include: age; gender and sexuality; social and economic class (education, job, family position); exceptionality (giftedness, differently-abled, health); geographic location (rural, suburban, urban, as well as north, south, east, west, or central); religion; political status; language; ethnicity (the aspect most people concentrate on when they think about culture); and racial designation (Ballengee-Morris & Striedieck, 1997; Banks & Banks, 1995; Gollnick & Chinn, 1993; Sleeter & Grant, 1988; Stuhr, 1995). While gender, religion, or language can influence one's culture, they are not the same as culture. For example, race and culture are not the same, although we constantly observe instances where people misuse these terms. We share age, gender, sexuality, social, economic, class, exceptionality, geographic location, religion, political status, language, ethnicity, and racial aspects of our personal cultural identity with different social groups. We are often greatly influenced by the national culture(s) in which these groups exist as well as the interplay of the political process. A person's existence and participation within groups are often the basis for positions of power and acts of discrimination. The various aspects of a person's cultural identity are in transition and dynamic—always changing. Recognizing our own sociocultural identity and our biases makes it easier to understand the multifaceted cultural identities of others.

We all have culture because we all live and exist within social groups. How we live our lives is influenced by aspects of our personal sociocultural identity as lived within a particular nation or nations and influenced by global issues.

High school is a tenuous time for identity development. Teenagers slowly develop a separation from caregivers' influence and sometimes move quickly into a rejection of both parents and community. This rejection focuses predominantly on outside influences, relationships with parents and guardians, and developing connections with friends. Countercultures, cliques, visual differences, social gatherings, music, and language are physical manifestations of this rejection through the development of new cultures. A person's existence and participation within cultural groups is often the basis for positions of power and acts of discrimination, which is prevalent during teens' lives. Being a member of the "in" crowd during high school does not guarantee a lifetime of happiness and success, but it does provide power, privilege, and possibilities. That is why purchasing or obtaining "in" items becomes so important during these years. With a means to obtain objects, high school students stand a greater possibility of being accepted or included into the desired sociocultural group. One example is when teens participate in shoplifting. According to the KidsHealth website (http://www.kidshealth.org/), teens make up the majority of shoplifters. Although there are many reasons that explain why this is the case, the first and major reason is that teens want to "fit in" or "be cool,"—they want to have the same clothes/items that their classmates own.

The development of identity is a fluid process, influenced by multiple structures, views, and people. Stuart Hall (1992), a well-known identity expert did not reject the notion of identity, but rather he problematized it by stating, "Identity is a structured representation which only achieves

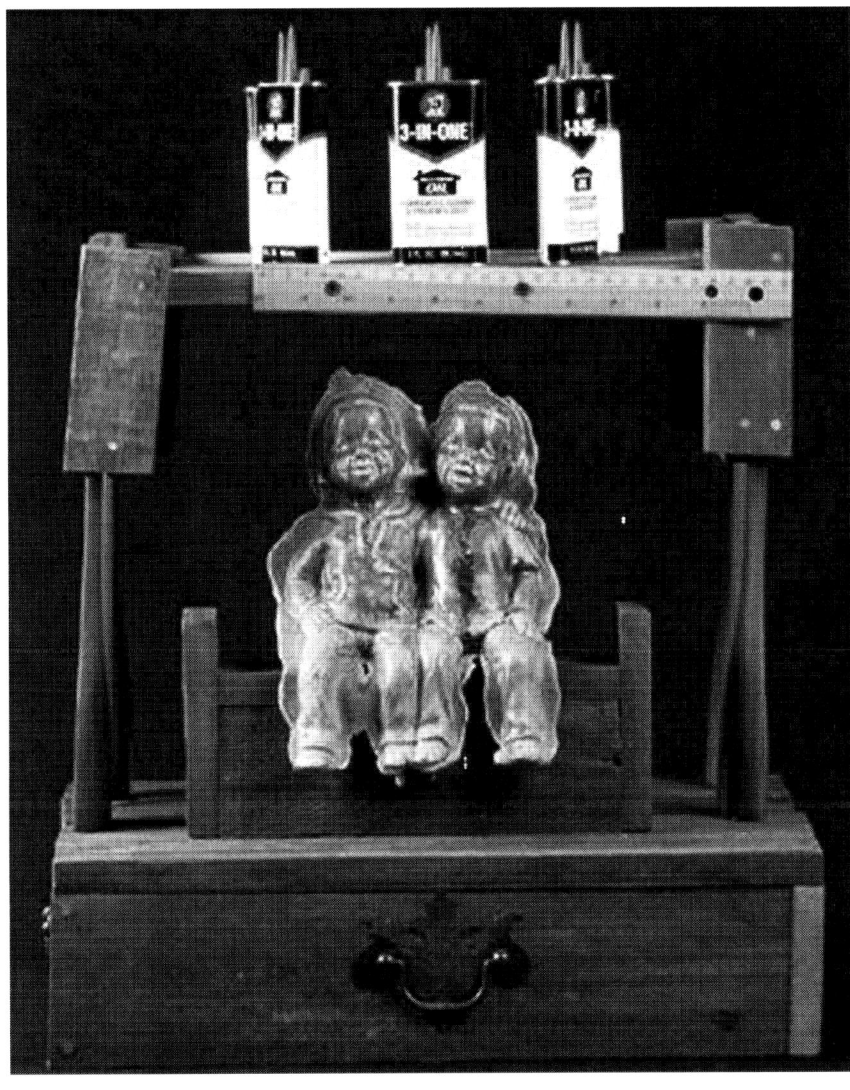

Figure 6.4 B. Stephen Carpenter, II, *Monkey Myth* (2001). Two ceramic figures of young African American boys with their arms around each other, sit smiling on a bench, covered with a honey gold-colored glaze. Above their heads rests a crown of 3-in-1 Oil cans perched on a pergola made of scraps of wood, wooden spoons, and wooden rulers. The title and imagery in this assemblage reference a variety of cultural and racial myths and connections, among which include the derogatory term "porch monkey," the signifying monkey, the trickster, manual labor, and the racist belief that black people were monkeys rather than humans.

its positive through the narrow eye of the negative. It has to go through the eye of the needle of the 'other' before it can construct itself" (p. 273). Identities are complex, complicated, and often ambiguous. Hall (2004) argued that identity construction needs to back away from relating only to difference– "I am different"– to a place of production– "I am this today." The need to reflect about histories of privilege, prejudice, oppression, and actions all become part of this dialogue that often guides people to question their beliefs and behaviors. (See Figure 6.4) Social identity development models (Tatum, 1992) provide guidelines for students to explore who they and others are and concepts related to identity such as sexism, privilege, heterosexism, and racism. These models also provide ways in which reactions such as anger, pain, and resistance evoked by the exploration of identity can be understood.

Since the teen stage is primarily about the construction of identity, schools and subject areas should offer multiple ways to include identity development, diversity issues such as class, oppression, and privilege, and the management of identity development. Just as the creation of an identity is normal for teens, knowledge of how to manage conflict with peers, parents, guardians, and other adults is also important.

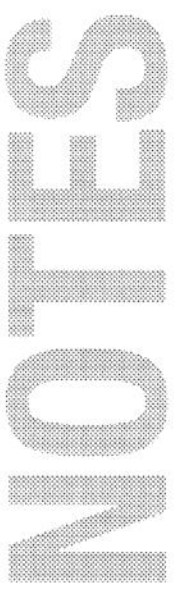

Students need to recognize their own biases and those of others in order to see the connections among concepts such as power, wealth, and injustice.

Pedagogy, Culture, and Student-driven Curricula

Rebecca anxiously walked through the hallway on the first day of school and hoped that no one noticed her.[4] As she walked down the corridor, she noticed that she had not encountered another Asian American person at her school so far that day. She felt that everyone was staring and pointing at her, but she knew this was not true. Jeremy passed a girl with dark black hair and hoped that she did not notice his limp. Since he started school, students called Jeremy names. He thought that this move to a new school would be a chance for a fresh start and to be accepted as part of a group. Shakeer closed his locker and found himself in the same stride with a white boy who seemed to have a slight limp. Shakeer smiled at him and hoped that his smile would be returned. Since his father's transfer to this country from Lebanon, he had not made any friends but had witnessed many curious glances after several students overheard him respond to a geography teacher's questions last month about what it was like to live in the United States as a Muslim. Fear of prejudice ran through the minds of each of these students. All of these students and many more, each with her or his own story and fears, sit down in your classroom everyday.

The needs and issues of all students including those like Rebecca, Jeremy, and Shakeer should drive curricula. Diversity cannot be explored through a top-down approach. When possible, teachers should move students conceptually and physically outside of the classroom and link with real-world communities, issues, and problems. Critical investigation is not without threat or danger; thus, teachers must be empathetic, practical, and cautious as they carefully create mentally and physically safe environments in which learning can occur. Teachers must help their students examine biases that lead to prejudice, discrimination, and colonialism through the exploration of history, current social issues, and visual culture. Students need to recognize their own biases and those of others in order to see the connections among concepts such as power, wealth, and injustice. The concepts of justice and equitable opportunities for all are important goals. To help students in understanding these goals, teachers must create opportunities in which learners can actively participate in and experience these concepts firsthand. We strongly believe that the examination and production of visual culture imagery and objects that lead to and end in an understanding of justice and the complexities of social, political, and economic relations are valuable goals for a meaningful interdisciplinary art education. [See Chapter 5 for an example.]

The strongest statement that can be made about a critical approach to interdisciplinary art education through cultural diversity is evidenced in how a teacher models the inclusion of social justice in the classroom. Imagine that Rebecca, Jeremy, and Shakeer are sitting down in your classroom. Who else do you see? How will you address these students—do you see them? Will you attempt to have your students get to know each other beyond their names, what they own or do not own? Or what they look like? Many books and resources for classroom use address activities that explore cultural identity and social justice. One general activity that we find helps students understand the complexities of culture is through an exercise in which individual students graph their cultural perceptions using concentric circles. The middle circle represents the person and the outer circles represent the visual or internal definers that a student uses to describe her or himself. Students draw these circles and record their ideas in their journals. Once this 5 minute activity is completed, the teacher asks the students to silently inventory what was written and more importantly what was not written—did not feel safe to share, not appropriate at this time, etc.—and why. Then, in small groups, students talk about the identifiers they are comfortable sharing with each other. Will this activity by itself stop prejudice and hate crimes? No. But the activity does begin to put a face to the name and helps students to move beyond assumptions about other people simply based on external appearances. One can only change prejudice and hate though the heart. Through such stories, names, and knowledge, in combination with explicit modeling from teachers of how and what to teach to foster inclusion, we can all work toward positive change and a better society. (See Figure 6.5).

We believe that high school classrooms should serve as a microcosm for a democratic society, in which students are cared for and learn how to care for each other. Caring about students and enabling them to care about others helps them to create environments in which they may feel safe and secure. This approach also affords a means for them to become hopeful, joyful, kind, visionary, and affirming. Granted, these are qualities that we all would like to see employed in the larger democratic society. Our goal is for students to envision themselves as people who value their own lives, embrace integrity, and are advocates and activists for justice. This vision for our students is possible only through an interdisciplinary/integrated curriculum that is conceptually connected to the cultural lives of students as well as academically demanding.

Content/Curricula and Aesthetics

Multicultural education has been around for over thirty years and yet the content and philosophy in many classrooms stems primarily from Western, European, or Euro-American aesthetics. This situation is especially

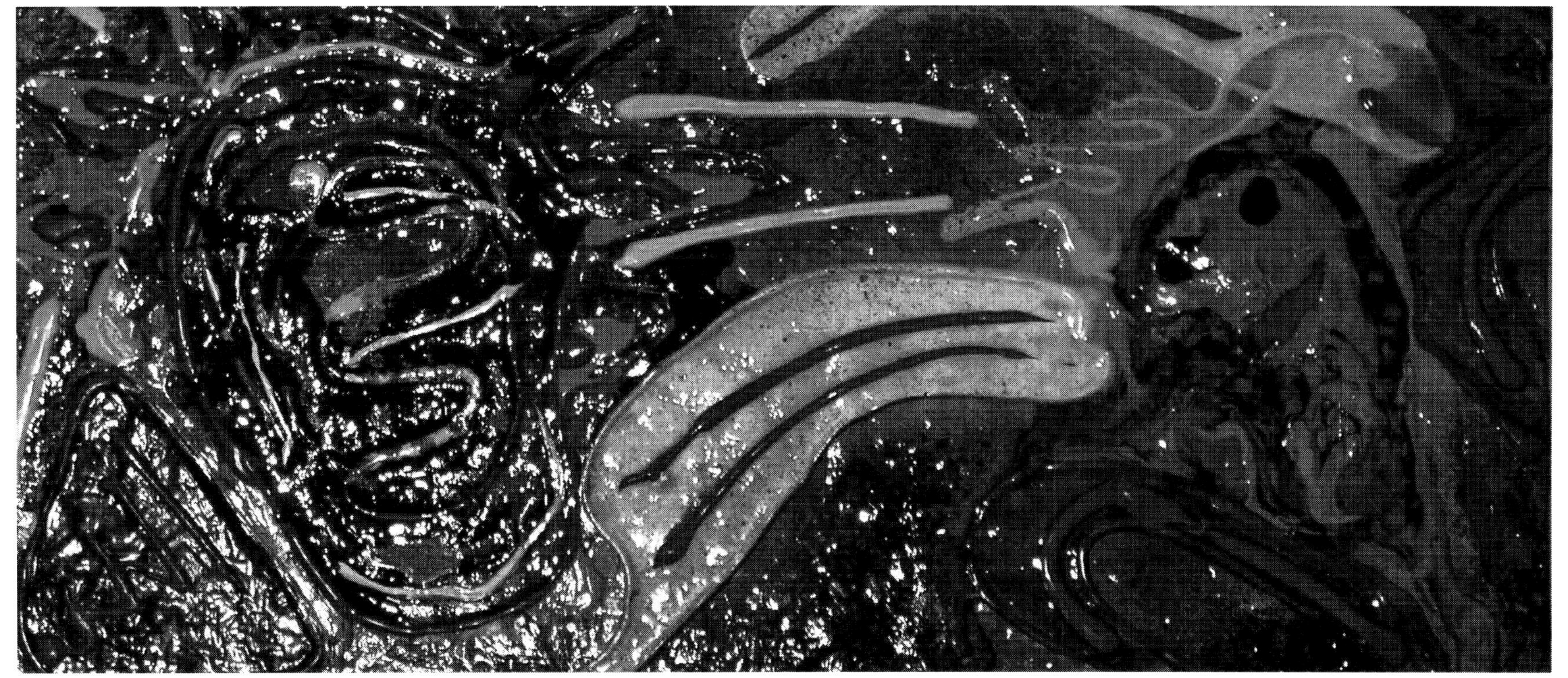

Figure 6.5 Jeremy Meisel's piece *Lady Liberty Turn Her Head* explores fear turned into hatred and manifested into global cultural violence. Permission of the artist.

> **Instead of looking at art works for what you want them to be—examples of color, composition, etc. Look at them and, most importantly, respect them for what they are—sacred objects, symbols, and important narratives.**

problematic when students try to transfer what they learn in school to other areas of their lives. If problems are not recognized and discussed, confusion or misinterpretation may occur. An example of this problem is illustrated when one explores Native art through a Euro/Euro-American/Western-based aesthetics. Native art is closely related to cultural identity, development of identity, and collective identity constructed around religion and nature, yet Western-based aesthetics is in direct conflict with these ideals (Ballengee-Morris, 2006). This situation means that design principles, elements, and possibly individual uniqueness are inadequate components or lenses through which to view or explore certain Native images. **[See Chapter 9 for more discussion concerning the elements and principles of design.]** American Indian aesthetics include the exploration of symbols, clans, traditions, rituals, and the transformation of those components as an individual relates to the group or clan. To view and explore Native American art without using Native aesthetics could be viewed by some Native Americans as a racist act and, at the very least disrespectful. Misinterpretations result when Native art or any other art is viewed according to standards not of that culture.

As Brent Wilson (1996) suggested, misinterpreting works of art is like telling lies. For example, before the teacher and students researched Faith Ringgold's *Sunflower Quilt at Arles* in the perspective unit discussed in Chapter 1 of this book, they may have judged her work to be primitive or naïve. In fact, comparing works of art from different cultures can be extremely problematic when adhering to a Western definition of aesthetics. Linear perspective originated as a Western or European compositional device.[5] As we learned from Ringgold's work, linear perspective was, in fact, antithetical to some Nigerian artistic traditions that represented value according to compositional size and placement of figures in works of art. Similarly, the art of Ancient Egyptians reflected human value through gesture and proportion. Figures representing royalty and status were represented in the classic (unrealistic) stance—legs, arms, and head facing west from a full frontal torso.

When exploring art from cultures other than our own, it is important too to be wary of taking or using cultural artistic devices in student artmaking activities. Materials, objects, signs, and symbols are often considered culturally sacred. Just as it is not appropriate for our students to use human blood to create a painting in high school, it would be considered blasphemous to have them recreate a Sioux warrior shield or a slave quilt. There is a fine line between role-play and trivial pretension. (See artist and activist Charlene Teters at http://www.charleneteters.com/).

So, how do you approach the art of cultures other than your own in the interdisciplinary high school art classroom? We believe that culturally specific works of art should be approached carefully, informatively, sensitively, and socially responsibly. Do not look at art works for what you want them to be—examples of color, composition, etc. Look at them and, most importantly, respect them for what they are—sacred objects, symbols, and important narratives.

Visual Culture and Diversity

Visual culture deals with images from mass media such as television, movies, music videos, computer technology, advertisements, magazines, and newspapers. These images create meaning and a vision of life for our students (Taylor & Ballengee-Morris, 2003). Using known images engages students and teachers with a sense of educational purpose and connects students with their communities and the visual culture that surrounds them (Bigelow, Harvey, Karp & Miller, 2001). The development of knowledge and skills is a process that depends on the connection of information that students already have with new information. This implies that the processes are inquiry based and intimately involve the students and teachers in the curriculum through which new knowledge and skills are developed. Visual culture education has evolved to include the interaction of issues of difference, the behaviors attached to shifting and conflicting social and cultural perspectives, and the relationships and roles of images as representations of personal and cultural norms, beliefs, and expectations. Visual culture is the term given to the changed and expanded understanding of art that is reflected in art education through the propagation and frequency of visual images and artifacts and the significance in people's everyday lives. **[See Chapter 9 for more information about visual culture.]**

Visual culture provides a stimulating component to interdisciplinary/integrated curriculum because it poses images and objects that characterize complexity, ambiguity, contradiction, paradox, and multiple perspectives. In such a curriculum, students learn that real life is messy and that issues are not solved by always having the "right" answer. **[See Chapter 8.]** Solutions involve discussion, compromise, negotiation, arbitration, mediation, cultural sensitivity, and openness to community resources, multiple representations, and re-representations (Bigelow, Harvey, Karp & Miller, 2001; Daniel, 2002).

One example of a visual culture controversy involving diversity issues is the Abercrombie and Fitch clothing line and resulting advertisements. This company markets t-shirts and other sportswear to young adults, teens, and children. Abercrombie and Fitch has been subjected to several legal cases against them: one for advocating underage drinking with a t-shirt "Drinking 101," racism with t-shirts that stated "Wong Brothers Laundry Service: Two Wongs can make it white," and "It's all relative in West Virginia;" provocative marketing to teen girls and boys that promotes active sexual behavior. Based on these discrimatory practices, the courts ordered Abercrombie and Fitch to make a $40 million settlement to Latinos, African Americans, Asian Americans, and women. The exploration of the visual culture and business strategies of companies such as Abercrombie and Fitch, teen participation, and the effect of collusion demonstrates the power of visual representation, marketing, and images.

Cultural Identity: Don't Box Me In

At the beginning of this chapter, we shared the story of Hope, a young woman who in her words was "white, American Indian, black, and other stuff, too." She did not want to be boxed-in to one cultural identity. At the high school level, the exploration of identities is an essential stage of development, and is often influenced by the values of social and visual/pop culture identities. Often self-hate, denial, and other negative feelings come into play. As students have cognitive developmental stages, they also have cultural identity developmental stages that are clearly mapped out in the educational curricula of many states. In the first two years in the primary grades, curricula tend to deal with individuals and the family. In the next year students typically explore community and by the fourth grade, students are exploring their state. Most often, fifth grade students deal with national culture and sixth graders begin to explore global connections. Middle and high school curricula typically expect students to repeat this cycle. Students are "me" oriented in their elementary years, they become "self aware" during middle school and are influenced by peers and the need to belong to peer groups. For this reason the repetition of these areas of study in the curriculum are not only more

NOTES:

Figure 6.6 Christine Ballengee-Morris. *Self Identities*—a sculpture that explores complexities and conflicts within identity development

theoretically complex, but are viewed from an "I" to a "we" perspective in high school. Using this evolution as the foundation of curriculum development, areas, issues, and themes in an interdisciplinary/integrated curriculum should be closely related to the cultural developmental stages and state curricula guidelines for strong discipline connections and relevant educational opportunities. As teachers know, students continually try on identities and are influenced by peers and society during their high school years.

These various aspects of cultural identity are transitional and dynamic—identities are always changing. (See Figure 6.6) Exploration and critique of the complexities, ambiguities, sociocultural identity, and biases of who and what we are make it easier to understand the multifaceted cultural identities of others. The relationship and intersections of personal, cultural, national, and global identities are continually being constructed and reconstructed in accordance with the current political climate. High school students may be introduced to the nuances and significance of national culture through classes, community involvement, visual culture, mass media (i.e. television, radio, newspapers, telephones, and faxes), and computer technology (i.e. E-mail, World Wide Web).

The personal, national, and global aspects of culture make up a fluid, dynamic mesh of an individual's cultural identity. During high school, many cultural components such as community involvement, visual culture, resistance, and complacency may become established. Therefore, it is vitally important for educators to recognize the educational opportunities during this time of development in the lives of adolescents. Often, cultural identity development during high school is viewed from a health, conduct, and negative perspective for some educators and administrators. Such a perspective far too often trivially translates into the imposition of new dress codes rather than "facing up" to the cultural implications of such behavior.

In secondary schools, an exploration of the cultural complexity of students and issues of power associated with social affiliations and aspects of personal, national, and global cultural identity(ies) is not only compelling but necessary. Teachers can stimulate and establish relevancy when they connect sociocultural issues, questions, problems, concepts, or topics to students' personal cultural identity level. Since visual culture plays such a prevalent part in students' lives and is connected to multiple disciplines across the curriculum, we support the use of visual culture as a meaningful way to explore cultural development and issues.

A Unit on Cultural Citizenship

We offer the following example of an integrated unit format for exploring diversity. In this unit of study the key concept is "Building Blocks of Cultural Citizenship: Story Telling and Accounting for Culture." Three essential questions that come from these key concepts are: How is national culture politically driven by narrative, and how are the accounts similar or different for those in the majority and members of minority communities? How does narrative shape national culture and frame our identities? Why do we wish some stories were no longer told? Using the Abercrombie and Fitch t-shirt and slogan, "It's All Relative in West Virginia," students will explore the historical, political, and cultural contexts of the saying. They may discover that the incestuous reference in the t-shirt slogan refers to the term hillbilly. Following the Civil War in the United States, scouts found that the Appalachian region had a great deal of natural resources that were in demand due to the Industrial Revolution. Several government and big business tactics were used to gather those resources as cheaply as possible such as land grabs—"a dollar an acre." Stereotyping the people as lazy, incestuous, and moonshine drinkers promoted a negative view of the citizens from West Virginia and became an acceptable excuse for cheating them. The term, hillbilly became popular in literature as a means to describe the victims of such behavior and soon found its way into the visual industry of movies, cartoons, and advertisements. Businesses such as coal mining and logging changed the economic-ways of the region. Native West Virginia people became dependent on these companies due to broken promises and exploitation of their land and ways of living. The resulting poverty became intertwined into the definition of their culture. At the same time, the crafts and music of the region were being "discovered" and the artisans were described as simple and quaint folks (Ballengee-Morris, 1995, 1997, 2000a, 2000b). Their creations were sold for large amounts of money made payable to the broker not the craftspeople. Years of oppression and misinterpretation created complex understandings and misunderstandings of the culture and the region. To this day, the hillbilly image is often celebrated by some members of the culture, while others detest it. Those not of the culture may not even know of the multiple contexts in which that image is held; therefore, many people have no idea of the derogatory nature of the term or image. But to say the least, marketing the term to new generations of teens through such television shows as *The Beverly Hillbillies* and more recently the Abercrombie and Fitch t-shirts, continues the racist representation. Through a critical study of this controversy, students may explore the use and practices by the National/Federal government in their region and critically analyze the purposes of marketing. They may compare their research with other accounts or versions of the stories they discover. They may look at the ways such stories shape their identities and arts and describe steps that they believe should be taken to help build understanding of others and counteract the impact of negative marketing.

Conclusion

Social issues, culture, art, and multiple perspectives can be integrated into all subjects within relations of power, history and experiences (Weiler, 1988). Curricular change is a transformative process, not a static product that can be singularly adopted. The new paradigm of interdisciplinary and integrated curriculum is one that continues to change and evolve through comfortable transitional steps that have significant personal and societal meanings. Teachers must remember that what they teach, how they teach it, what they connect to the lives of students and to world events should help students question, challenge, and change their world. Just as the students' cultures are always changing, their needs, understandings, and learning styles also change. Our job as educators is to provide opportunities and learning experiences that enable them connect to their culture, learning, thinking, and the world.

Suggested Discussion Questions and Activities

1. Read one of the readings in the following list and write a critical analysis based on key points, curricula, and/or pedagogical suggestions, and outcomes. Consider questions such as what is being advocated and is it reasonable? What are the stated outcomes and what outcomes are not stated? Why is diversity necessary or is it necessary, why or why not?

Banks, J. A. & Banks, C. M. (1989). *Multicultural education: Issues*

and perspectives. Newton, MA: Allyn and Bacon.
Campbell, D. & Campbell D. D. (1999). *Choosing democracy: A practical guide to multicultural education*. Upper Saddle River, NJ: Prentice Hall.
Freire, P. (1997). *The pedagogy of hope*. New York: Continuum.
Gay, G. (2000). *Culturally responsive teaching: Theory, research and practice*. New York: Teachers College.
Hill, M. (Ed.). (1997). *Whiteness: A critical reader*. New York: New York University Press.
Linton, S. (1998). *Claiming disability*. New York and London: New York University Press.
Loffreda, B. (2000). *Losing Matt Shepard: Life and politics in the aftermath of anti-gay murder*. New York: Columbia University Press.
Takaki, R. (1993). A different mirror. Boston: Little, Brown.
Timpson, W. M.; Canetto, S. S.; Borrayo, E. & Yang, R. (2003). *Teaching diversity: Challenges and complexities, identities and integrity*. Madison, WI: Atwood Publishing.

2. Your school celebrates United Black World Month in February. You have saved all of the Black artists for that month to talk about and you have developed some projects that will demonstrate for your students why the Harlem Renaissance occurred. As a white teacher, you have learned so much from developing this unit and yet some of your African American colleagues seem to be critical. Why do you think they are critical of your plans? What would you do and why?

3. Your school has little ethnic/racial diversity. Often you find yourself challenged by this situation but you continue to work to include diversity issues in your teaching. One day during one of your lectures, you ask the only Native American student in the school to give the Native perspective about Navajo rugs. The student bows her head, mutters something, and then is silent. The other students in the classroom begin to be disruptive with "Hollywood" whoop cries and you respond with comments to them about their rude behavior. What should you have asked yourself and the student? Why would the student have acted that way? What should you have done?

4. Over past years, there has been a sudden increase in hate crimes that demonstrate homophobic beliefs. More often you hear students telling homophobic jokes or making negative comments about their classmates in the hallways between classes and in the cafeteria. What can you do in your classroom to discuss the issues of homophobia and hate crimes? What do you need to do to be able to discuss this in your classroom?

Notes

[1] This story is based on experiences of authors Christine Ballengee-Morris and B. Stephen Carpenter.
[2] This image is the center image of a triptych of photographs by James Luna. The left image depicts the artist in profile facing toward the right with long, black hair. The right image depicts the artist in profile facing to the left with short black hair and a moustache. http://www.msu.edu/user/sullivan/Photo/PhotoJLunaHalfMex.html
[3] This story is based on author Christine Ballengee-Morris' experiences.
[4] This story was constructed from authors Christine Ballengee-Morris, B. Stephen Carpenter, and Pamela G. Taylor's teaching experiences.
[5] The quality of portraying real figures in real pictorial space is a distinct Renaissance characteristic. The invention of mathematical rules for correct perspective came from Filippo Brunelleschi (1377-1446), who trained as a goldsmith. Brunelleschi made at least two paintings in correct perspective, but is best remembered for designing buildings and over-seeing the building works. The first written account of a method of constructing pictures in correct perspective is found in a treatise written by Leon Battista Alberti (1404 – 1472). The first version, written in Latin, was entitled De pictura (On painting), and was Alberti's effort to relate the development of painting in Florence with his own theories on art. An Italian version (Della pittura) appeared the following year, and was dedicated to Filippo Brunelleschi. (South, 2006).

Chapter 7
COMMUNITY & WORLD CONNECTIONS THROUGH SERVICE-LEARNING

"Our relevance as human beings can be seen in the meaning of our acts as artists" (Estella Conwill Májozo, 1995, p. 88).

Beginning in September 1987, artist Dominique Mazeud walked the riverbed of the Sante Fe River on the seventeenth day of each month for many years. She brought large garbage bags and with the help of friends and local community volunteers, worked to collect and remove the trash that littered the river's beaches and shallow areas. Originally, Mazeud's performance art was designed as a political catalyst for environmental awareness. Over the years, however, this process has changed to become more of a reverent ritual and connection with the river. She wrote, "All rivers have currents and are connected. . . People function in the same way. One way to activate these currents is through ritual. Rituals are icons of connections, they are the art of our lives" (Lacy, 1995, p. 263).

Ritual was intricately connected to the garden art of artist and Floyd (Virginia) High School art teacher Catherine Pauley.[1] On a small piece of land adjacent to a county landfill and overlooking Buffalo Mountain, Ms. Pauley created a healing and meditation garden. Begun as a way to connect with something living and growing and a way to rehabilitate her arm after surgery, Pauley welcomed anyone to participate and/or visit her rocky cliff art

Service-learning Models — Community — Reflection

Experiential Learning — Social Issues

garden. The work was homage to the power of nature.

Artists create works of art to understand, explain, provoke, examine, explore, and expand the meaning of what it is to be human. As postmodern artist and critic Suzi Gablik (1991) put it, "art can transcend the distanced formality of aesthetics and dare to respond to the cries of the world" (p. 100). Some artists such as Mazeud and Pauley took personal action to make those imagined possibilities a reality through their art and collaborative artmaking process.

Over the course of two years, students at a low-income housing development's after-school program in a rural western Virginia town near a liberal arts university, worked with local university students to create a flower and vegetable garden in a vacant lot on the property. The goal was to take a neglected area that had become a haven for dangerous activities in the community and create a living and growing testament to the power and growth of the community.[2] A similar garden project involved students at Cedar Shoals High School in Athens, Georgia in the research, design, soil preparation and planting of a rain garden to transform a muddy detention basin into a garden classroom.[3] The project continued with the creation of an outdoor sculpture, a butterfly garden, and walking paths.

Creating an art garden takes time. Like Robert Irwin's Getty Garden (Weschler, 2002) [See Chapter 2], the Kraus Campo Garden at Carnegie Mellon University (Bochner, 2004), the Spiral Garden at Bloorview MacMillan Children's Centre in Toronto, Canada (Bloorview Macmillan, 2005) and the Art Gardens of Pittsburgh, Pennsylvania (2002), garden art projects involve multiple ways of thinking, knowing, and doing. It takes many hours and many hands to create a community art garden. Thinking, planning, and working together take time. Even then the best-laid plans must be re-evaluated and altered due to weather, environmental, and/or development changes. We use the art garden as an example in this chapter for the ways that collaborative community service or art-based service-learning can become a catalyst for collaborative interdisciplinary learning.

A Closer Look at Service-Learning[4]

Service-learning involves students, teachers, and professionals actively working to use what they are learning in their formal study to help others and to make a difference in the world. But the idea of service in education is not new. Some of the first colleges and universities in America were founded on the idea of service. University service programs were established during the Great Depression in the 1930s. The call to service grew over the years and in 1990 President George H. W. Bush signed the National Community Service Act. Later, President Bill Clinton proposed legislation to expand opportunities for all Americans to serve their communities and to earn awards for their own education (Archer, 2000) establishing AmeriCorps through what is now the Corporation for National Service (VA COOL, 2000).

Much service-learning theory is derived from John Dewey's (1938/1963) theory of experience. Dewey explained that all genuine education came about through connective experience involving observation, knowledge, and judgment with each being highly subjective, personal, and experiential. He said, "the essential point is that education grows through a process of social intelligence" (Dewey 1938/1963, p. 72). Dewey's pragmatism, or as he called it, instrumentalism, was "the theory of education as deliberately conducted practice, and education as such, was not a means of living, but was identical with the operation of living a life" (Saltmarsh, 1996, p. 14). For Dewey, pedagogy connected practice and theory or "action and doing on one hand and knowledge and understanding on the other" (1932, p. 107). Dewey's idea that education should be linked to life and living can be linked to the service-learning goal of connecting theory to practice.

A number of service-learning programs are initiated through AmeriCorps and funded through grants from the Corporation for National Service. Such programs assist service partnerships among communities, schools, and institutions of higher education. At the state level, service-learning initiatives vary in scale and context. Students at De Soto Middle School in De Soto, Wisconsin extend their study of English and composition through service to local nursing homes. They visit local nursing homes, hold a Folk Fair with retired community members, and complete a book of stories that results from interviews they conduct with local nursing home residents (Wisconsin Department of Public Instruction, 2002). A ninth grade technology education class in Talbot County, Maryland creates public service announcements addressing specific community needs they identify after conducting research (Maryland Student Service Alliance, 2002). High school art students in Christiansburg, Virginia teach area middle and elementary students ceramics. They also design and create exciting art installations to inspire young students at the Christiansburg Public Library's summer reading program. Science students take part in the county's annual "Bloomin' and Broomin'" day, cleaning roadsides and planting flowers. Virginia Tech University students assisted in the establishment of an African American museum in a refurbished historically black public high school. These students augment programs in local schools through distance learning, and work with area businesses and schools on computer web page design (Service Learning Center at VA Tech, 2000). Shenandoah College in Winchester, Virginia students work with Spanish farm laborers to translate and complete work-forms and applications or provide home health care to shut-ins (Lesman & Morrow, 1999). Alicia Swackhamer, Forest Hills High School art teacher in Monroe, North Carolina, reported that her students created collages for a local hospital as a result of studying artist Romare Bearden. [Teacher contribution.] Her students took a tour of the hospital and learned how the many aspects of the facility and the people who work there were essential to the care and well-being of patients and families. Borrowing collage techniques and the idea of collage as a multi-dimensional representation of the world from artist Romare Bearden, the students worked together to make interdisciplinary visual metaphors that represented the mission of the hospital facility. Elisa Gargarella, Assistant Professor of Art Education at the University of Akron, Ohio facilitates a month-long summer art education residency program for high school students who are economically challenged.

One main goal of service-learning is to instill an enduring sense of civic responsibility.

> Each summer, 10-15 high school students are selected to work with a resident professional artist to create a public art installation for a partner organization in our community (Akron, Ohio). Arts LIFT is a service learning program that helps students develop a connection to their local and broader communities through the vehicle of art making. Each project aligns with the mission of our selected partner organizations that are all dedicated to the preservation of our local natural and cultural resources (email communication, December 28, 2005). [Teacher contribution.]

Art teacher Chris Kitzmiller reports that her students published two books for their high school and Apache Reservation community on diabetes and alcoholism along with literacy books and note cards in Cibecue, Arizona.

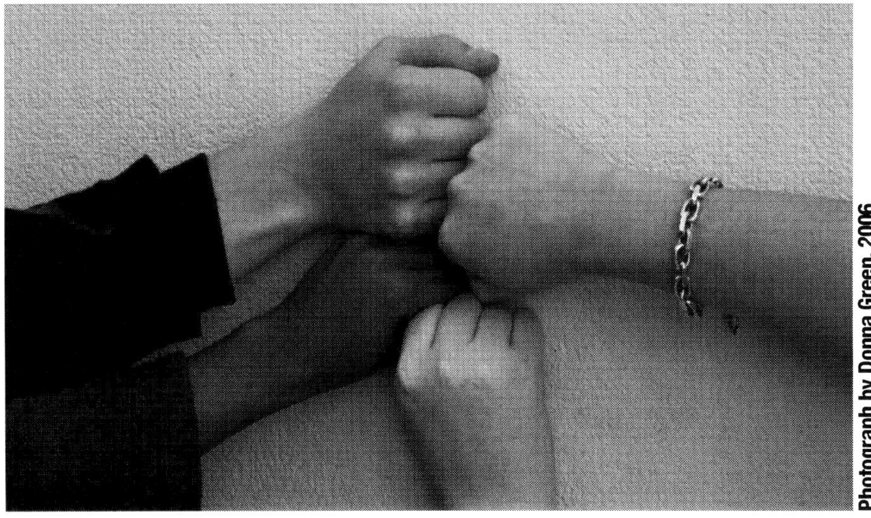

Photograph by Donna Green, 2006

[Teacher contribution.] Lincoln, Nebraska art educator Tara Adams also involves her students in working together to make change in their school community. [Teacher contribution.]

> We have discussions about: how things which happen in our community affect us, our lives now and in the future. In the beginning of the unit we discuss purposes of varied and political artists' works. We watch political documentaries, read poetry by such poets as Alice Walker and talk about action and reaction. We think about our "school community" and discuss ways to find our voice inside its campus confines. We discuss how art has the power to not only help people think but to feel emotion as well. The students document parts of their school that they would like change incorporating reasoning and planning strategies. Students consult and collaborate with other classes to make a final plan. Students complete the project and document it with writings, photographs, and a letter to the editor of the local newspaper. (personal communication, September 12, 2004).

Nationally, a number of schools, community agencies, and arts organizations work with the Empty Bowls Project (Taylor, 2002b). Each year, thousands of people create handmade ceramic bowls for the project. Essentially, the project is a simple meal of soup served at a fundraiser for local food banks and other agencies that provide nourishment for the hungry. Along with the meal, guests witness educational performances and exhibitions. At the end of the event, guests keep their bowl as a reminder of the empty bowls in the world (Empty Bowls, 2000). (See Figures 7.1, 7.2, and 7.3) Lifeworks Foundation in Nashville, Tennessee works diligently with Tennessee artists to promote Empty Bowls projects across the state. In addition, singer and songwriter Quinn Loggins freely gives his song to be used at any Empty Bowls event (see http://www.quinnloggins.com).

Models of Service-learning[5]

Because one single approach or way of working is rarely utilized in the real world of education, we present the following six basic models of service-learning (Heffernan, 2001) with the understanding that they will be used in a variety of connective ways and in varying de-

Figure 7.1. Brenda Stein, *In the Beginning*, turned wooden bowl created for Empty Bowls in Nashville, Tennessee. Permission of the artist.

grees. This list is not a set of absolutes but offers examples of ways that service learning can be identified, planned, and enacted. Following this list, we further illustrate these models and make connections to the art garden project we mentioned earlier in this chapter.

• Pure service-learning courses are not connected to any specific discipline or course. They are, however designed to compliment and connect to a variety of subjects through the act of service itself. Typically, pure service-learning courses link students to service organizations or agencies according to interest and need. In this model, it is the responsibility of the student to make the appropriate curricular connections through reflection, journal writing, and independent conference and study with specific discipline teachers or experts.

• Discipline-based service-learning is specific and explicit to a particular course or class. Service activities are formulated and practiced-based on the appropriateness of the experience to the standards and curriculum design of a specific discipline.

• Problem-based service-learning surrounds efforts that address and meet the needs of a particular issue in the community. The need is approached as a problem to be examined critically through reflection, identification of multiple solutions, and authentic connections with the learning that is going on in the class.

• Capstone service-learning courses involve students in the final course in a sequence of study or program. Students in a capstone service-learning course typically practice and critically reflect upon the ways that their own learning in that specific discipline is applicable.

• Service internships involve more time and are typically agency-centered. In these internships, the student's reflection activities are related more toward the service than toward a particular class or discipline.

• Action research requires students to conduct research activities as well as service. As in any action research project, students involved with action research report on actual practice. However, when critical reflection is a part of action research, as is the case in service-learning, the researcher works to effectively change that practice through direct links and connections between their learning in the classroom and their service work in the field.

Figure 7.2. Claudia Lee. *Empty Bowl.* Handmade paper. Permission of the artist.

At Empty Bowls fund-raising events, guests keep their handmade bowls as reminders of the empty bowls of the world's hungry.

The garden project at the low-income housing development's after-school program discussed at the beginning of this chapter began as a way for the community and service-learning students to create a flower and vegetable garden in a vacant lot on the property. Service-learning students already working in the after-school program as well as new students and volunteers took part in the project. Some of the students gained community service hours for extra credit in a class and had no specific course or content requirements for their work. Their experience was classified as pure service-learning. Granted, these, as all of the other service-learning students, used knowledge and skills they learned in their high school classes. However, in the case of the pure service-learning students, the connections with their class learning had to be recognized by the students independently. On the other hand, discipline-based service-learners from an earth science class worked with the garden project by using the skills and information they learned and included these connections in their formal and informal reflection practice.

The garden project is also an example of problem-based service-learning. The students and the community worked together to address the need to transform a crime-ridden and dangerous vacant area, adjacent to the housing complex, into a symbol of growth, beauty, and sustenance that could encourage community stewardship. The garden project involved advanced art service-learning students in their capstone course, the last art class they took before graduating. Childcare service-learning students used the project as part of their internship or practice-teaching hours. We see this project as a form of action research as evidenced by this and other accounts, as well as through further reflection in such courses as social studies and art for social justice.

Service-learning supports many approaches, disciplines, ideas, and ways of working with and for communities and the world. Teachers and curriculum designers must remember that there are some very basic and essential qualities that distinguish service as a learning activity and/or characterizes learning as a form of service.[6]

Authentic service-learning is discipline and/or interdisciplinary-based. Teachers and students work with and for their communities through various projects. These projects are often referred to as community service projects and are approached as separate somehow from the curriculum. Though learning undoubtedly occurs during such projects, if the service is directly related to the curriculum, as in the case of service-learning, then the experience becomes more of a way for students to connect theory to practice while they also increase their civic and citizenship skills (Howard, 1993).

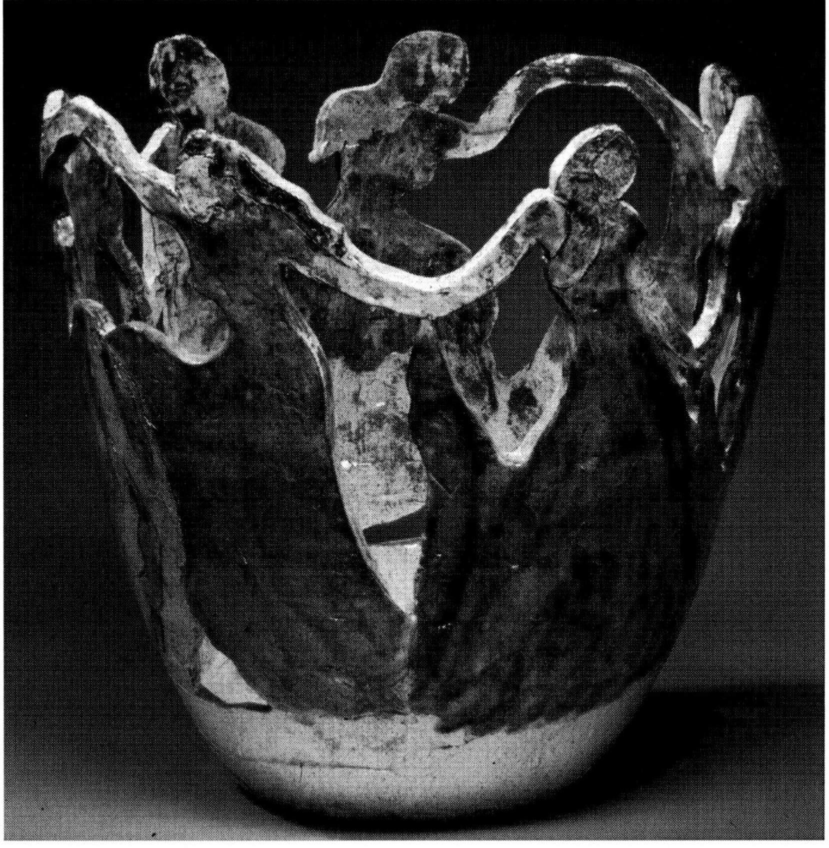

Figure 7.3. Donna Rizzo. Dancing Circle. 18 x 18 inches ceramic bowl created for Empty Bowls in Nashville, Tennessee. Permission of the artist.

Just as in any interdisciplinary approach, the service-learning project must be designed to enhance the curriculum and standards of all subjects/classes involved. Such could be the case if the garden project was classified as an interdisciplinary art and science service-learning project. Art and science students might work with the young children in the program to create a flower and vegetable garden. Back in their classrooms, service-learning students could study the work of artists such as Joseph Norman (Taylor, 2002a & 2003), Lynne Hull[7] and Mierle Laderman Ukeles.[8] They may relate the work of these artists to their study of sustainable growth in the environment in their ecology classes and/or to their study of plant species in their earth science classes. Working with each other as well as the staff and young children at the after school program, with proper approval high school students could design and create a garden. Earth science or botany students might germinate and grow the seedlings in their classes while art students continue their design study through the creation of permanent garden sculptures.

The service aspect of authentic service-learning must be based upon a communally recognized need. That need cannot be dictated. It must be agreed upon by the agency or group as well as the teacher or class involved in the service-learning experience. "Doing with" rather than "doing for" encourages an empowering form of service and promotes advocacy over charity. The after-school program at the low-income housing development in Virginia approached the school and the teachers in their quest for a garden project. The teachers then worked with each other and their students to create a socially and academically meaningful experience.

Another example of such a communally recognized need involved a Christiansburg (Virginia) High School art class in the creation of an installation for the summer reading program at a local public library. After the public library contacted the school, the art teacher worked with the world history teacher and the public library staff. Together they created meaningful and authentic art study and making activities that met curricular guidelines while at the same time provided an exhibition/installation that enhanced the library's Ancient Egyptian reading program.

The keystone of service-learning pedagogy is reflection. (See Figure 7.4) Through reflective discourse, writing, exhibition, and critique, service-learning students, teachers, and community participants must be actively involved in constant production and re-production of the service project. With mutual goals in mind, service-learners and community partners work and learn together in the same way that students learn from as well as teach their teachers (Freire, 1970/1994). Probably the most crucial aspect of service-learning theory is this reciprocity. Constant and meaningful exchanges between all parties involved are critical for mutual respect of everyone's values, needs and expectations. Such reflection promotes interdisciplinarity because it requires verbal communication as well as writing in a variety of forms. Reflection may take place immediately following each activity on-site, in the following class period in the classroom, through written commentary submitted as a paper or in electronic form such as e-mail, listservs, or online discussion bulletin boards. Reflection in the form of exhibition may involve the display of art made by students in response to the experience along with informative text panels, photographs, and

NOTES:

Ideally, service programs help students (and others) identify, acknowledge, strengthen, and utilize their voices to help bring about positive and lasting social justice.

implications for future projects at the school or the service site. Some teachers include reflection activities on tests either through discussion or specific questions that include examples from the service activity. For example, an earth science quiz or test following the garden project might include such questions or tasks as: Explain the formation of strong (covalent) chemical bonds between the atoms of carbon-containing (organic) molecules in the seedlings you nurtured for the garden project.[9] What factors involved in our service-learning garden project affect environmental quality?[10]

The garden project may be included in an art quiz, test or criticism activity through specific art-related questions such as: In what ways was the service-learning garden project similar to the goals and ways of working[11] involved in Merle Laderman Ukeles' *Methanogenesis*?[12] (Risatti, 1994). What design-specific elements came into play when we planned the garden? How and why were these plans altered as the garden matured? Visual reflections in the form of sketchbooks, digital photograph diaries, and/or specific artmaking assignments that accompany written text and formal criticism sessions are also excellent catalysts for disciplinary and interdisciplinary reflective thinking.

Committed involvement is essential to successful and meaningful service-learning programs. Like gardens that require continuous care, service-learning projects need to be either ongoing or touted as weekly, monthly or yearly events to promote trust among all parties involved. Single, isolated service acts rarely promote sustained social and civic responsibility. Trust is another important component to successful service learning projects because when a project is seen as continual, it can be changed and modified according to the needs of all involved. For example, although community children and adults continue to sustain the garden art project, every semester a new group of university service-learning students join in to work as well as provide inspiration for such seasonal tasks as planning, preparing soil, planting, harvesting, and winter clearing. Such continuous, committed involvement can take the form of yearly events as well as service activities that are integrated into the standard curriculum. Another such example involves high school business classes that require internships in local businesses. Such internships could become service-learning projects with the inclusion of a reflection component. Students would also need to perform these internships at non-profit organizations or other organizations that express specific need and whose aims are geared toward social and cultural opportunities.

Obviously if these sites are art-related such as a museum or local arts organization, the arts connection is clear. However, since many business internships involve promotional work such as the design and creation of visual materials, natural links to art and art class could also be made. And if for example, a local ballet company can count on the fact that their high school interns will assist with the bookkeeping and promotional materials for their yearly production of the *Nutcracker* ballet, the school, the teachers and the students are more likely to form a committed, lasting connection.

One main goal of service-learning is to instill an enduring sense of civic responsibility in students. The hope is that once students experience the self-satisfaction that comes from cooperative service, they will make a conscious effort to involve themselves in their future communities long after they graduate. Ideally, service programs "help students (and others) identify, acknowledge, strengthen, and utilize their voices to help bring about positive and lasting social justice. Yes, of course, work to alleviate the symptoms of injustice but work to understand the causes of the injustice and strive to alleviate those causes as well" (Empty Bowls, 2000). The Empty Bowls project is an excellent example of interdisciplinary service-learning. Begun in 1990 in a Michigan high school art class, Empty Bowls[13] is a unique and nationally recognized fundraising project that helps to alleviate hunger. Because the project involves various activities and components such as the making of ceramic bowls, soup, promotion, research, presentation, publications, fund raising, and service, interdisciplinary connections like art, science, food service, math, accounting, graphic design, English, social studies, theatre, and music are required. In addition, foreign language students may be involved in the translation of promotional materials in bilingual communities. The Empty Bowls project provides students with an almost immediate sense of accomplishment because of its fundraising focus. In addition, because local soup kitchens and food banks are in constant need for funds, many students will continue to participate in these yearly events long after they complete their schooling. Besides the yearly Empty Bowls event, students and teachers alike work with the soup kitchens and food banks through volunteer services, exhibition of art, assistance in promotional materials, food acquisition, and educational materials. When service-learning projects are meaningful across disciplinary boundaries and provide students with opportunities to reflect on the difference their participation truly makes in the lives of others, they are more likely to continue living with a sense of civic responsibility.

> **The service aspect of authentic service-learning must be based upon a communally recognized need. That need cannot be dictated. . . . "Doing with" rather than "doing for" encourages an empowering form of service and promotes advocacy over charity.**

Discussion Questions and Activities

1. What does the following quotation mean and to what essential characteristic of service-learning does it relate?

> Using community as a laboratory rather than working with the community on jointly useful projects may stunt the development of partnerships that offer continuous benefits to both parties (Eyler & Giles 1999, p. 179).

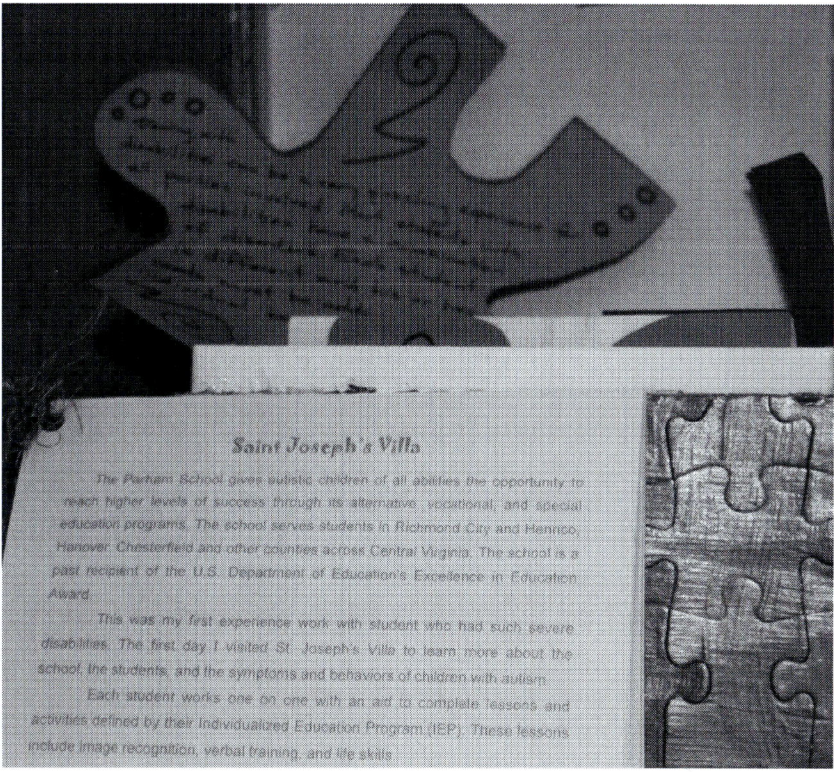

Figure 7.4. Art students in Donna Green's classes create visual reflection projects in response to their service-learning work with students with special needs. This student work used the puzzle metaphor to illustrate the ways her service-learning experience became an integral piece to her teacher training. Photograph by Donna Green, 2006.

2. In Chapter 5, we discussed some of the ways that art class is used as a school resource for creating signs, brochures, sharing supplies, and so forth. How could these and/or other ways in which the art class is considered a resource, become instead an interdisciplinary service-learning opportunity in the school or in the community?

3. In what kinds of community service have you been involved? How could you translate that experience into a service-learning?

4. Research your home community and create a list of agencies and contacts that could become service-learning partners in your future art teaching.

5. Read Chapter 2 in Carol Jeffers (2005) book *Spheres of Possibility*. Referring to Jeffer's discussion of Andrew Furco's (2001) visual metaphor of service-learning, construct a three-dimensional sphere using colors and text to represent the interdisciplinary possibilities of a service-learning experience.

Notes

[1]Author Pamela G. Taylor worked closely with Catherine Pauley during the years 1997-2001. Ms. Pauley was a dedicated cooperating teacher for Radford University art student teachers as well as the leader of the local Floyd County Church House Gallery.

[2]This project took place during author Pamela G. Taylor's experiences working with the Beans and Rice, Inc. program and Radford University 1997-2001.

[3]Author Pamela G. Taylor became aware of this project during her time at the University of Georgia 2001-2004.

[4]Much of this section of the chapter comes directly from author Pamela G. Taylor's (2002d). Service learning as postmodern art and pedagogy. *Studies in art education*. *43*(2), 124-140.

[5]Much of this section of the chapter comes directly from author Pamela G. Taylor's (2004a). Service-learning Basics. *NAEA Advisory*, Spring.

[6]See authors' Pamela G. Taylor and Christine Ballengee-Morris (2004). Service-learning; A language of "we." *Art Education*, *57*(5), 6-12.

[7]Hull creates new animal and ornithological habitats in natural spaces that have been damaged by human encroachment. See http://www.eco-art.org/

[8]Ukeles was a keynote speaker at the National Art Education Association Conference in New York on March 15, 2001. For more information regarding her art see http://www.feldmangallery.com/pages/artistsrffa/artuke01.html, http://www.astc.org/exhibitions/rotten/ukeles.htm, http://www.newyorkartworld.com/reviews/ukeles.html and Lacy, 1995.

[9]This activity is linked directly with the National Science Content Standard C: As a result of their activities in grades 9-12, all students should develop understanding of matter, energy and organization in living systems (NSES, 1995).

[10]This question is directly related to the National Science Content Standard F: As a result of their activities in grades 9-12, all students should develop understanding of environmental quality (NSES, 1995).

[11]This question directly relates to proficiency levels in the National Visual Arts Standards # 3. Choosing and evaluating a range of subject matter, symbols, and ideas reflect on how artworks differ visually, spatially, temporally, and functionally, and describe how these are related to history and culture and # 4. Understanding the visual arts in relation to history and cultures (analyze and interpret artworks for relationships among form, context, purposes, and critical models showing understanding of the work of critics, historians, aestheticians, and artists).

[12]Merle Laderman Ukeles' *Methanogenesis* involved a community of artists working together to create works of art that mimicked the natural methane producing effects of anareobic bacteria while at the same time warned against its harmful effects in landfills. In the process they issued a challenge to themselves and the viewers of the art to work together as a community toward a sustainable and ecological way of living (Taylor, 1997). Also see http://raykass.com/html/ukeles01.html

[13]The Imagine /RENDER Group, a 501(c)3 organization, promotes the project (http://www.emptybowls.net).

Chapter 8
PROMOTING INTERDISCIPLINARITY THROUGH DIGITAL COMPUTER TECHNOLOGY

> I'll start something in the computer hypertext and then I discover something really exciting that may have influenced the artist in a particular work or something. So, I go to a book or the Internet and research that and then invariably I find something else that takes me in an entirely different direction. I just don't feel that I can know enough or find out enough anymore. It just doesn't stop! It's changed everything!
> (Elizabeth, personal communication, April 22, 1999).

Imagine an art classroom filled with students who thirst for knowledge so much that they feel they just cannot know or learn enough? Picture in your mind what such an art class might look like. There are tables, of course piled high with artmaking supplies, where students busily work while they laugh and talk. These same students work on paintings and drawings at their easels and throw pots on the potter's wheel. The floor is covered with clay dust, charcoal, scraps of paper and splashes of paint. As usual, piles of artworks are on the teacher's desk, supplies and "in process" works of art spill from shelves, and in every corner of the room are stacks of materials that some day may become a part of someone's art.

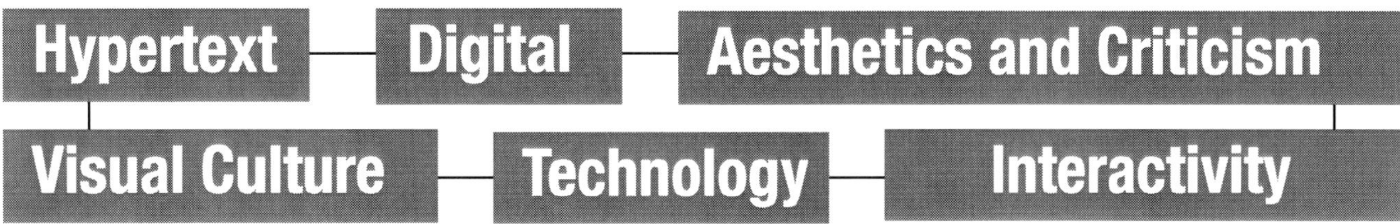

As you are probably thinking, this picture is no different from any typical art classroom scene. So, what if we added to this image several students working on laptop and desktop computers? What if some students take pictures of themselves and each other with digital cameras and their cell phones while others view clips of movies and television shows on their iPods™ ? Is there still no difference? And, what if the students working on the computers are not just working on their own individual works of art but instead are going back and forth between the computers, the tables, the easels, and other areas of the room continually throughout the class period? Suppose that all students at some point are writing, researching, making images of their art, and chatting online with artists, students in other schools, and with each other.

They move back and forth among the computers and the bookshelves. They read, search and flip pages in books and then race back to the computer to type or scan a new discovery. One student runs to a world map, points to a country as she looks closely to find a particular place, and then traces her finger back to another location, perhaps one closer to home. Another student squints as he reads the label on a bottle of paint while a classmate brings a book to a table to show something to the students working there. More students seem to be watching television, but a closer look reveals that they are actually recording video clips. As they make comments about what they see, they also point at the screen and leave thousands of fingerprints.

In addition to this scene, imagine what the students in this art classroom are saying:

> Look at this painting by Magritte (*The False Mirror*). It's like the eye in the Madonna video[1] that keeps coming back. Let's put it in the web.
>
> Hey, this book says that Magritte was part of a group of artists known as Surrealists. They worked in Europe during the 1920s and rebelled against conventional art values. That's exactly what Madonna is always doing, rebelling! So, if we do a work on rebellion, then we could use say, paint to make a sculpture or something. How would we do that?
>
> Yea, I found a website that said surrealists based their art on memories, feelings, and dreams and used fantastic visual techniques like levitation, scale, and juxtaposition. Remember the ways that horn kept floating around in the video. We could do a painting and float it or something.
>
> I found an artist in this book who makes things float (*David Smith: Cubi 5 and the Floating Cube*).
>
> Speaking of floating, what do you think about using some of these ideas for the art club's homecoming parade float? Like film clips with art and symbols that stick out from the car?

Now imagine what these students are doing on the computer. As part of their study of one of pop singer Madonna's music videos, one student types information about artist Frida Kahlo in a writing space of the class's communal hypermedia web. In Kahlo's *Tree of Hope*, this student saw a disturbing similarity between the painting and images in the video of wounds on the backs of female figures. Linking the word "back" to another student's explanation of a similar Man Ray work, *Violin de Ingres* this student connects the communal web to his research on the influence of Ingres and his painting *Nude from the Back*. The student links another Kahlo work, the *Suicide of Dorothy* with a video clip of Madonna flying through the air. Another student pastes a scanned image of the Malawi dervishes' sama dance he found in an encyclopedia into the computer hypermedia project. (See Figure 8.2.). He types explanatory notes in another space on the computer. One of his classmates creates a link from that space to a clip from another video while he makes still another link to a space that contains his research about the artist Rene Magritte.

For several days, the students add stills and short excerpts from videos, song lyrics, and personal comments in their computer hypermedia web.(See Figure 8.1.). Throughout this process the teacher looks at, listens to, and reflects on what she sees. She thinks about the kinds of ideas that the students are researching and the directions that their links are taking. She reads the student work and follows the trails of thought represented in their links and explanations. She sees opportunities to engage her students in critical discussions or avenues for further discovery about the meanings of the connections they have made. The teacher even considers a change in the direction of her planned activities for the next week and the rest of the semester based on the enthusiasm of her students and the insightfulness of their connections, comments, discussions, and questions. In this process she learns from her students while they are simultaneously learning from her and each other.

Do you see a difference now? What is going on? The difference in our new scenario is that the flurry of activity is not limited to the making of art. Now, the activity also requires students to think about and link artworks with ideas and other realms of experience, both inside and outside the art classroom through the magic of digital computer technology.

Digital computer technology may be relatively new and developing to some of us, but it is the only environment in which many people worldwide under the age of 25 know to be standard.[2] With a mouse or remote control in hand, many young people spend hours in front of their computer and television screens as image after image is thrown at them at the speed of light. This rapid, fast-paced way of seeing is routine for many contemporary young people. Unlike many adults, they are not amazed by the technology. It is simply part of their everyday existence. They want the things they see on their digital screens and they try to look like or act like specific characters and television stars. Their ideas about government, war, politics, interpersonal relationships, team dynamics, family values, religion, law and order are played out their television and computer screens. They fill their art with images and symbols from their favorite web sites or television commercials, cartoons, music videos, and films. They play computer games as they simultaneously watch television, chat online, and research with the help of search engines. They travel through time and space with cell phones, MP3 players, webcams, and laptops. These "digital kids" see and interpret many aspects of their lives through their digital lenses of experience.

Digital computer technology may be relatively new and developing to some of us, but it is the only environment in which many people worldwide under the age of 25 know to be standard.

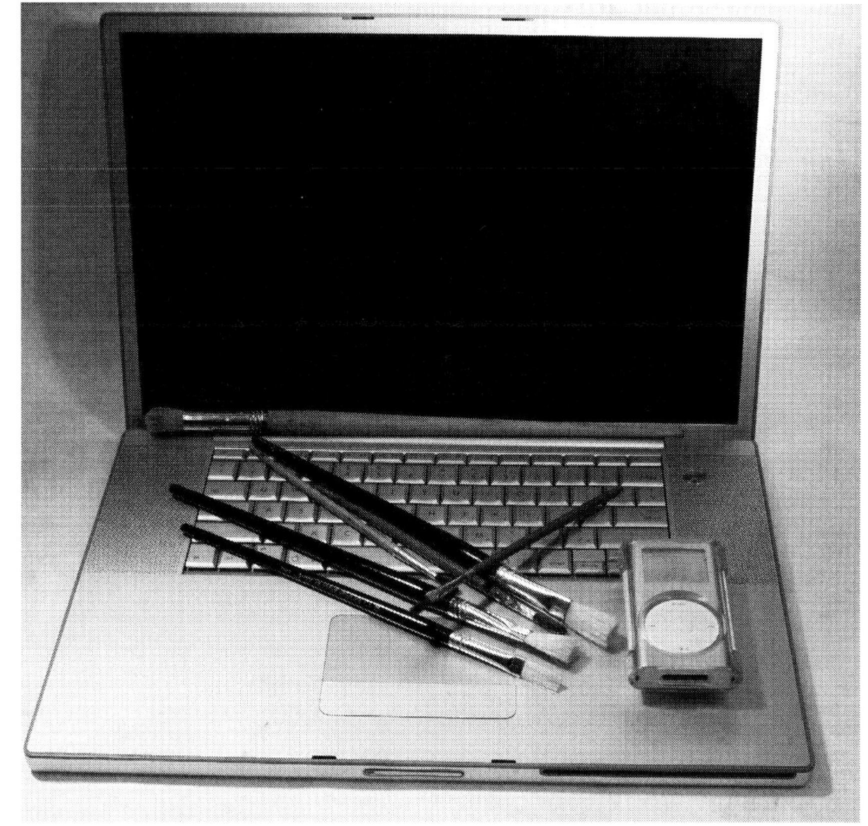

Photograph by Donna Green, 2006.

For us, hyperlinks embody the interdisciplinary nature and most provocative aspect of interactive computer technology and the Internet because of their connective and associative capabilities. A simple click of the mouse on a web page reveals another page with other links. We, like our students, know that the Internet is a collection of information and connections between many ideas, people, places, and machines.

In addition to a critical approach to the content and context of digital technology, in this chapter we look more closely at the ways that interdisciplinarity might be a natural way of working in and through interactive computer technology. We suggest varied instructional strategies for using digital computer technology (computer, television and film) as a catalyst for and record of connective thinking and provide exemplary models of art criticism activities, historical contextual connections, intertextual technique and media exploration. In addition, we offer strategies that employ such interactive computer technology as hypertext, digital film/movie making, web design and aesthetics, computer generated images and projects, and online teaching and learning.

Digitally-Mediated Interdisciplinary Curriculum

We believe that computers should be a means by which we teach and not an end to what we teach. In other words, the use of a computer should help students and teachers think about art, not simply make art. We recognize the fact that thinking and making in art curricula are not new concepts. In fact, art teachers and students think about art and make art every day. The question is how can students think about art and make art in ways that are more personally meaningful and relevant to their lives. That is, we are interested in ways in which students and teachers can use computer technology as an interdisciplinary means to think about, question, look, analyze, interpret, make, and understand the art and visual works they see, experience, and produce.

The activities, atmosphere, and dialogue described earlier in this chapter were not made possible simply because the students sat in front of the computers. The flurry of activity in

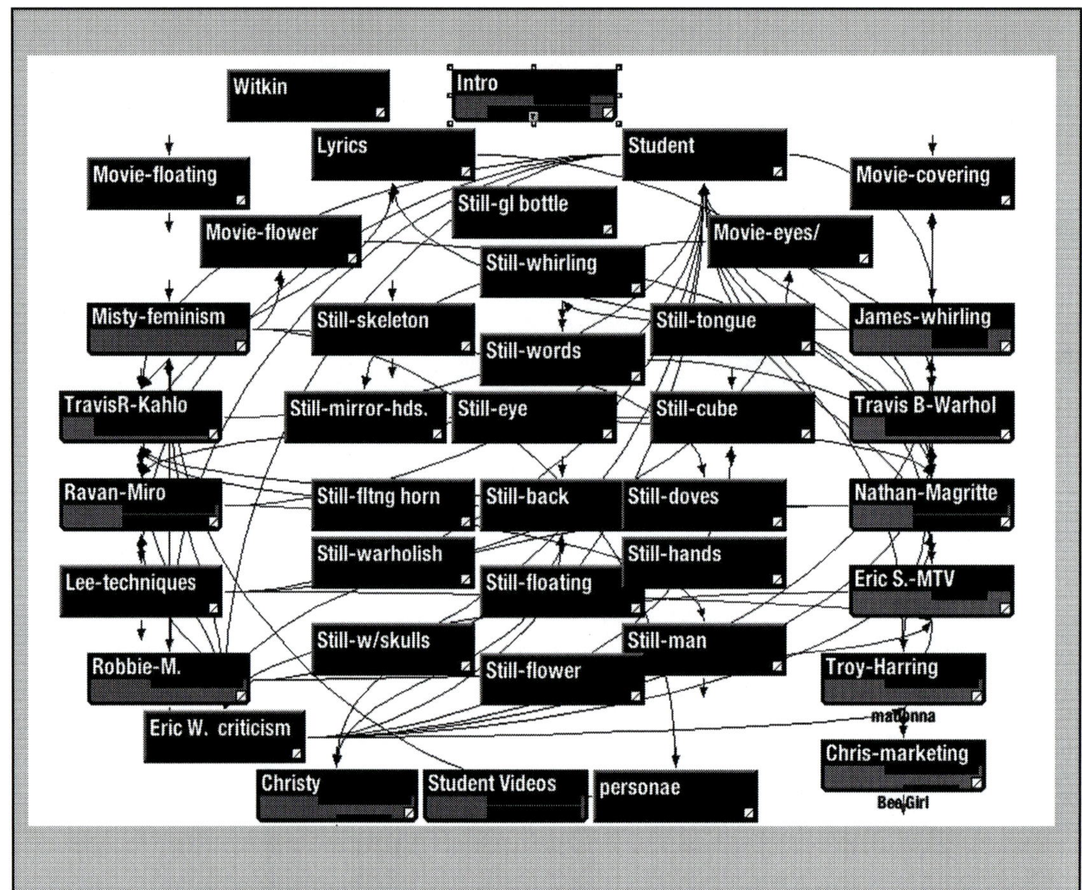

Figure 8.1. As part of a high school art class' study of one of pop singer Madonna's music videos, students conducted a wide variety of research and then placed and connected their discoveries in a hypertextual web on the computer.

that scenario is not typically associated with computer work. The students engaged in talk, research, discovery and collaboration both on and off the computer. They thought about what they were doing. They posed questions and searched for answers. They changed their minds. They worked with others to flesh out ideas and to connect the ways of working they learned about in class and their own research. They challenged each other in positive ways by asking meaningful questions about techniques, media, content, images, and ideas. And in the process they recognized and confronted connections between their lives in and outside of school. They were completely immersed in an interactive computer-mediated interdisciplinary curriculum.

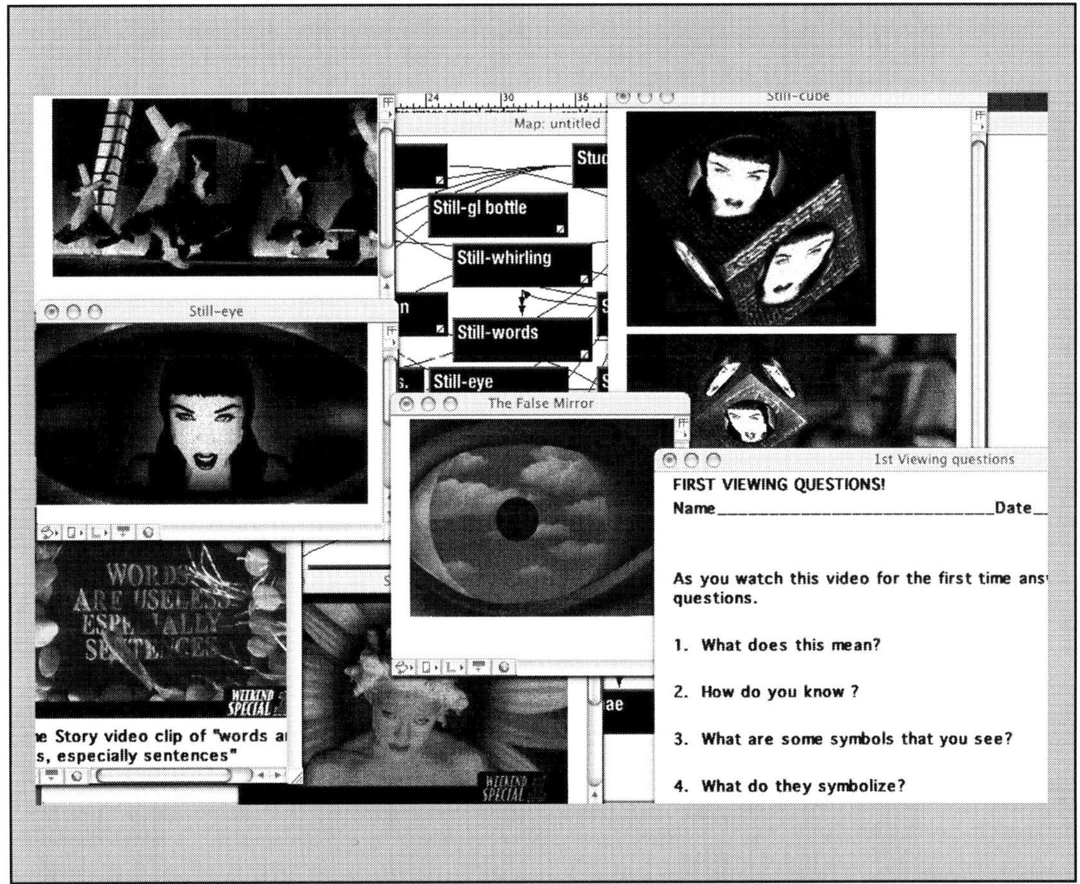

Figure 8.2. One student added an image of the Malawi dervishes' sama dance explaining that dizziness to them was a higher state of consciousness. He linked this space to another student's research of the artist Rene Magritte.

Don Wass (2004), an author and K-12 art teacher at Northwestern Regional School, Winsted, Connecticut reports that his students' use of virtual reality computer software sparked an interdisciplinary boom of activities and possibilities in the school. **[Teacher contribution.]** From the virtual reconstruction of a foundry operated during the American Revolution to the reconstruction of math and calculus concepts in virtual spaces, Mr. Wass and his students worked with a number of teachers in his school to create "3-D simulations of real or imagined environments using advanced software and design techniques" (p. 41).[3]

In Miami, Florida, Coral Reef Magnet School art teacher Colette Stemple uses computer software in her classes to experiment "with actions, paths, and filters to apply 'visual math' to their imaging projects" (Johnson 2004, p. 2). Both Ms. Stemple and Mr. Wass use computer software to visualize such math concepts as perspective. **[Teacher contribution.]** Mr. Wass explains, "What the brain tells the viewer about a form is not well represented on a flat piece of paper but it is very well represented in a 3D environment" (Wass in Johnson 2004, p. 3).

National Board Certified high school teacher Philip Conrad's "Leonardo da Vinci" interdisciplinary unit of instruction uses applied technology at Hamilton Wenham Regional High School in Hamilton, Massachusetts. Conrad maps the unit on the computer to reveal the intricate connections between: the lessons; the Massachusetts Curriculum Frameworks for Fine Arts and Science and Technology/Engineering; the evaluation criteria; field trips; and materials. For example, in an interdisciplinary problem about walking buoys and Water Walkers, Mr. Conrad's students design, construct and test a set of walking buoys modeled after the design sketched by Leonardo daVinci. They field-test their creations on Chebacco Lake. In a similar engineering design process, Mr. Conrad's students construct musical instruments, paint frescos, create emblems, and host a banquet. Although his students use computer technology in their research and design process, what is significant is the way that Mr. Conrad uses the computer to demonstrate the connective and interdisciplinary nature of this unit. **[Teacher contribution.]**

Author B. Stephen Carpenter II involves his art education students in the creation of digital movies based on big ideas and key concepts. Using digital video cameras and iMovie software, his students explore the use of images, sounds, and actions to examine and portray such ideas as power and identity. Essentially, his students create 2-minute concept commercials to "advertise" the units of instruction they design. In much the same way that abstracts serve journal articles and movie trailers (previews) function as promotions for full-length movies, Carpenter's students conceive of their concept commercials as abstracts or trailers for their units of instruction. His students design their curriculum units around interpretations and research of works of art or visual culture and must reference—either explicitly or implicitly—those works of art or visual culture in their digital movies.

In 2005, Ms. Anna Golden, an instructor at Virginia Commonwealth University (VCU) worked with art education pre-service majors at VCU to involve adolescents in the Blackwell Summer Art Program in the re-creation of their idea of home through digitally collaged images of interior and exterior places in their community. **[Teacher contribution.]** The Blackwell Summer Art Program was initiated in Richmond, Virginia in 1999 by Dr. Charles Bleick, current Associate Dean for Academic Affairs, VCU School of the Arts in Qatar. Bleick, community leaders, and university students developed the summer art program for young people

Computers and digital media should be a means by which we teach and not an end to what we teach. Students and teachers can use computer technology as an interdisciplinary means to think, question, look, analyze, interpret, make, and understand the art and visual works they see, experience, and produce.

whose families had been displaced during a housing redevelopment project in the city. Over the years, the program continued with varied community-driven art activities. Ms. Golden's experience reveals an increasing bonus of interdisciplinary instruction—the possibilities for connecting and combining technology, art, and service. **[See Chapter 7 for more information regarding service-learning.]**

Sue Raymond, art teacher at Horizon High School, Scottsdale, Arizona uses the World Wide Web to involve her ceramics students in reflective writing. **[Teacher contribution.]** Dan Springer, an art teacher at Dennis-Yarmouth Regional High School in South Yarmouth, Massachusetts reports that he draws upon history, sociology, psychology, and media literacy when teaching storyboarding in his Art in Advertising course. **[Teacher contribution.]** Using the World Wide Web as a showcase for student art as well as instructional tool is especially evident in Lance Wurst's art teaching at Perkins County Public Schools in Grant Nebraska. From sketchbook assignments to student finished art and lesson plans to pacing guides, Mr. Wurst reports that in his small western school, interdisciplinary and technological approaches are an integral part of his every day teaching. **[Teacher contribution.]**

Similarly, as an art teacher in Kentucky, artist photographer Wendy Ewald[4] began using photo technology with her students to engage them

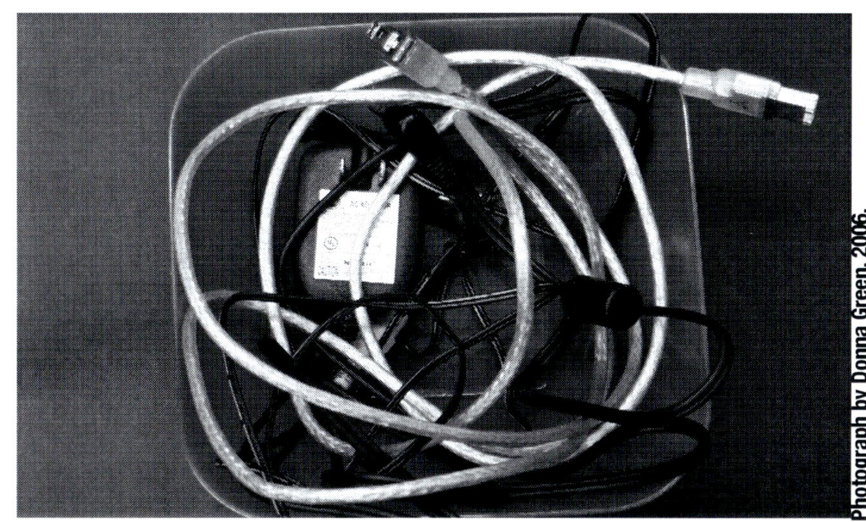

Photograph by Donna Green, 2006.

to imagine their dreams and fantasies. For over twenty-five years, Ewald has worked with children in Columbia, South Africa, Mexico, the Netherlands and many areas in the United States. Through her work, Ewald and the children examine critically how the world in which they live defines the way they view their hopes, dreams, and identities.

Visual Culture. Earlier in this chapter we described how a piece of digital visual culture—a music video—functioned as an important interdisciplinary catalyst in the art classroom. **(See Chapter 9 for more on visual culture.)** Many high school students are more acquainted with such visual culture as music videos, restaurant advertisements, video games, pages from fashion magazines, and recognizable billboards than they are with the works of art, literature, and music that they study in the classroom. The art classroom is a context in which knowledgeable discussions about visual culture can and should be nourished and explored in connection with the curricula of visual art and other school subjects (Carpenter, 2003; Barrett 2003a, Duncum 2001, Freedman 2003, Taylor & Ballengee-Morris, 2003). According to Henry Giroux (1992):

> If we view the world as a text, then literacy means engaging the full range of what is in the library (conventional notions of reading), the art gallery (the making and interpretation of art), and the street (popular culture and student experience) (p. 243).

In other words, educational goals should include visual and media literacy because both contain and embody elements of the text of our world in and out of the classroom. The International Visual Literacy Association (2005) defines visual literacy as:

> A group of vision competencies a human being can develop by seeing and at the same time having and integrating other sensory experiences. The development of these competencies is fundamental to normal human learning. When developed, they enable a visually literate person to discriminate and interpret the visual actions, objects, and/or symbols, natural or man-made, that are [encountered] in [the] environment. Through the creative use of these competencies, [we are] able to communicate with others. Through the appreciative use of these competencies, [we are] able to comprehend and enjoy the masterworks of visual communications (Fransecky & Debes, 1972, p. 7).

Media literacy as defined by The Center for Media Literacy (2003) is "the ability to communicate competently in all media forms,

NOTES:

Through a simple click of the computer mouse, teachers, students, parents and administrators can follow links among the activities going on in the art classroom and the curricular standards that drive them.

print and electronic, as well as to access, understand, analyze and evaluate the powerful images, words and sounds that make up our contemporary mass media culture" (para. 3). We believe that students must develop critical viewing practices of such popular and visual culture as television and learn how to deconstruct cultural identity representations in the mass media if they are to become informed citizens of this democracy (Taylor & MacDonald, 1996). **[Visual culture and interdisciplinary high school art is explored further in Chapter 9.]**

Hypertext. In his book *The Quiet Evolution*, Brent Wilson (1997) referred to the disciplines of aesthetics, criticism, art history, and art making as lenses. He presented diagrams of overlapping circles to represent the combined power of the discipline lenses to view and "transform the art object more effectively into an interpreted and understood work of art" (p. 97). He used the diagrams to holistically describe, plan, and assess a unit or lesson plan in relationship with a work of art. Wilson further illustrates the ways discipline lenses may be used in the creation and development of thematic units of instruction based on key and related works of art and a selection of student works surrounding the basic theme of the key work. Wilson (1997) placed clusters of circles that represented the discipline lenses on top of boxes that contained student work, studied works, student art writings, related art works including music, literature, theater, and related historical, critical and philosophical writings (p. 103-105).

Wilson's idea of lenses and diagrams was greatly informed by Marjorie Wilson's (1998) use of hypertext computer software in art criticism and art education curriculum design. A number of hypertext computer programs are available for this purpose and more are developed each year.[5] These computer programs basically operate under the same connective premise of providing readers and the writers (authors) the ability to create, organize, and rearrange notes or thoughts on the computer.

Each electronic note may contain a variety of contents and be linked electronically to other notes through hyperlinks. Because of the changeable nature of these notes, hypertext provides a site for continual redirection in the processes of thinking, interpreting, and critically responding. Such re-direction and reflection is possible through computer hypertext because of the ease with which the computer allows readers and writers to maneuver through past, present, and future thinking (Taylor & Carpenter, 2002; Carpenter & Taylor, 2003). Unlike note cards, books, or papers, computer hypertext enables the creation of visible links between thoughts, ideas, images, and parts of images. By following paths throughout a hypertext web, teachers and students can track the thinking process of the web's creator. And with a mere click of the mouse to gain access, readers can add challenges, questions, and comparisons directly in a note. Doing so opens multiple notes at one time rather than requiring the reader to shuffle through numerous pages of a book or scraps of paper. Basically, the World Wide Web on the Internet is hypertext as are varied and continually emerging software packages such as Inspiration™, Powerpoint™, Hypermedia™, Storyspace™, and Tinderbox™.

Unit and Lesson Planning. The process of planning lessons and units in an interdisciplinary curriculum mediated through such digital computer technology as hypertext involves the teacher in a connective form of research that is directly related to local, state and national curriculum standards. Because of the connective nature and powers of the computer to access information, teachers can make explicit connections easily between their plans and the standards. Typically, a lesson plan includes a space for listing the curriculum standards to be covered in the experience. A unit or lesson plan in a computer-mediated curriculum provides the teacher with a tool to link, as well as a way to see possible interdisciplinary and curricular connections. Through a simple click of the computer mouse, teachers, students, parents and administrators can follow links among the activities that are currently the focus of study in the art classroom and the curricular standards that drive them. We find that a computer-mediated curriculum typically exceeds the goals and standards for a given grade level or course, as students and teachers become more actively involved in not only what they are expected to learn, but how they learn and why.

In a computer-mediated curriculum, the teacher is involved in connective research and planning that is also personally meaningful. The teacher, in essence, practices that which she or he will teach, and learns through the reflective and discovery process of seeing and creating electronic links between, around, above and below works of art, criticism, art history, popular culture, and other realms of experience. The special powers of interactive computer technology both enable and require this kind of activity.

When teaching a computer-mediated curriculum, the teacher may or may not share her or his own hypertextual discovery process with students. However, the teacher's own process remains crucial to the development of the activities and instructional strategies that she or he will employ. For example, a teacher may discover a significant link between a work of art, a big idea or key concept, and a television commercial or other area of popular culture when creating a lesson or unit plan. He or she may create a link from this information to a video. In doing so, the teacher might also formulate an activity in which students research and present a form of popular culture that deals with the big idea of the unit and later explain symbolic or metaphorical links they have made between their and other's works of art. Obviously, such an open-ended activity results in uncertainty to say the least. Therefore, guidelines or criteria should be established in order to prevent students from violating school regulations. Such guidelines or criteria however, should not be too restrictive, as a basic premise of a computer-mediated curriculum is the encouragement of expansion, growth, exploration, and alternative paths of discovery. A computer-mediated curriculum celebrates what John Dewey called "flexible purposing" in which the point, objective or aim of the curriculum grows and takes shape through the contributions of all engaged in the process.[6] In other words, a computer-mediated curriculum leaves room for and encourages change and growth. **[See Chapter 2 for more information regarding lesson planning.]**

Unit Templates. Computer hypertextual templates might be used for unit planning, interpretation and criticism activities, aesthetics discovery activities and/or process portfolios. A unit plan template could include

Photograph by Donna Green, 2006.

Technology is not thought of as special anymore. . .It is simply part of our students' everyday experiences.

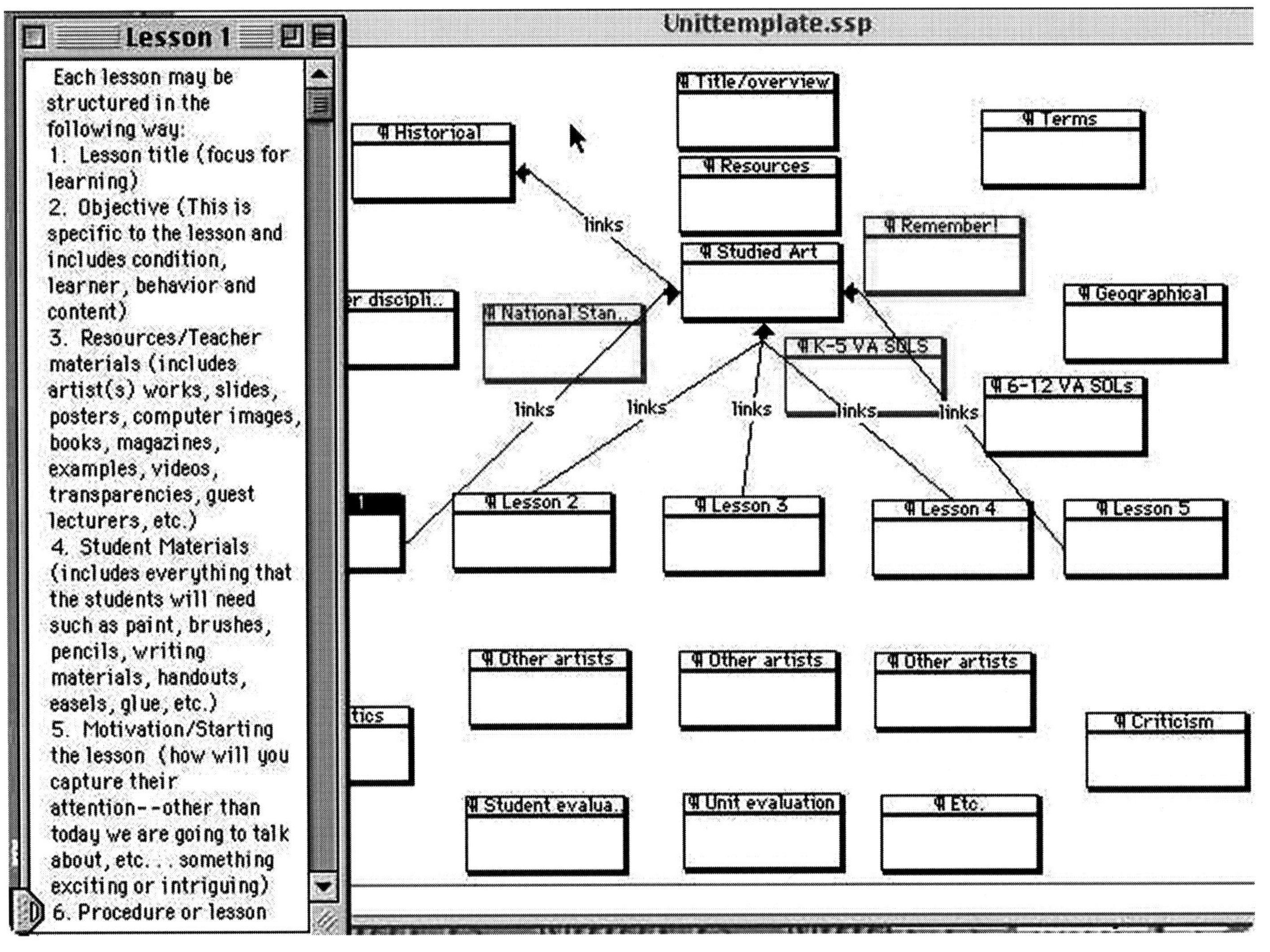

A basic premise of a computer-mediated curriculum is the encouragement of expansion, growth, exploration, and alternative paths of discovery.

Figure 8.3. A unit plan template could include spaces for the unit overview, individual lesson plans, state and district standards and resources such as maps, charts, World Wide Web links, vocabulary, critical writings, and information about artists, films, student examples, and procedural directions.

spaces for the unit overview, individual lesson plans, state and district standards and resources such as maps, charts, World Wide Web links, vocabulary, critical writings, and information about artists, films, student examples, and procedural directions. (See Figure 8.3.). Teachers can make links between the pieces of information they discover and ponder new ideas and connections among the information and activities that comprise their units of instruction.

As art education students and teachers move from one unit to the next, they should revisit questions, themes and information revealed through interpretations of works of art studied in previous units. Traditional linear approaches to curriculum design and interpretation make hypertextual reading seem awkward at first. When teachers and students develop and respond to essential questions at any time during the reading or writing of a hypertextual document, they learn to more easily navigate the complexities of their hypertextual units.[7] Collaborative activities, working from templates and student-designed models may assist the teaching of computer hypertextual unit formulation.

Interdisciplinary Art Criticism with Computers. Because a key goal of art criticism and interpretation is to literally construct meaning by building connections from a work of art to other texts, students should be encouraged to ask questions and offer uncertainty about the meanings of works of art.[8] In a computer-mediated learning environment, each student constructs interpretations based on his or her own frame of reference and sets of knowledge. By doing so, other students may comment upon, dispute and/or link information from one computer space to another. References should be established to the cultural contexts in which the work, the artist, the historians and the critics exist. Visible signs in the works can be actively and consciously translated into verbal and written interpretations. Initial interpretations can be combined with those of other students and sources to produce more complex interpretations and understandings. One student might view a work as a political statement, and another could emphasize the historical significance of the same work. With sufficient supporting information supplied by the students, each of these interpretations could be plausible and connected—technically, visually, and conceptually—in the computer.

Criticism templates. A template for interpretation and criticism activities may contain spaces that ask specific questions about the art, artist and student. Several spaces may ask questions about the artwork, the artist's life, the political and social world at the time the art was created, or other influential works of art by the artist and others. Other spaces may feature questions about the student's own life, art, or world-view, along with a space for an imaginary or real dialogue with the artist and/or the other students in the class. An aesthetics discovery template may contain spaces for works of art along with quotations by aestheticians, philosophical questions and puzzles. A process portfolio template would contain spaces for images of works of art studied in the class, and student journal entries and art. Once the students complete the prescribed templates and have a base from which to work, they are free to create links and add other personally important information.

Assessment. Assessment of such computer constructions is an ongoing and collaborative process. As learning is an active mental process, authentic assessment should take place throughout the student's art learning and artmaking experiences. Through the use of hypertextual computer environments, teachers and students read and/or present information from their web constructions on a regular basis. Teachers comment and make suggestions directly in the student hypertext webs. By doing this, the suggestions and comments offered by the teacher become relevant to the students' ongoing processes, and encourage reflection and revision. **[See Chapter 10 for more information regarding assessment and evaluation strategies and examples.]**

The Internet. Access to the Internet is now available in most schools and offers a wide range of interdisciplinary possibilities for

NOTES:

Photograph by Donna Green, 2006.

teachers and high school students. Millions of images of artworks, film clips, news stories, techniques, styles, books, videos, and television shows are available online. In addition, a number of World Wide Web sites are devoted to student activities that range from interactive stories to astronomy and scavenger hunts to printmaking. World Wide Web search engines also offer possibilities for teachers and students to discover resource information in the form of web sites, images, local information, university research sites, and more. We find these sources of information and images particularly useful when we create visual presentations and lesson and unit plans. Many museum web sites offer visitors amazing three-dimensional views of galleries and art. Podcasts[9] are available on many museum sites as well enabling museum visitors to download information about particular exhibits in lieu of traditional docent-led tours or on-site listening devices. Non-sanctioned exhibition podcasts are becoming increasingly popular. Though often derisive, such rogue podcasts are appealing to high school students because of their anarchist bent. Of course, caution must be taken when utilizing podcasts, we believe that we should not ignore the fresh and relevant views they sometimes provide.

The Internet also offers many possibilities for interdisciplinary collaboration such as discussions among students and teachers around the world or simply across the room through e-mail, discussion boards, chatrooms, and online course software. With the availability of data storage sites, students and teachers may access collaboratively created files and web pages. Online interactive educational games are available or students and teachers can create their own. Teachers and students can use Internet access to download films, access videos, conduct research, and purchase virtually anything from a building to a gardening tool at any time from anywhere.

Along with the exciting opportunities for communication and information retrieval on the Internet come important safety and ethical issues. Some schools establish firewalls or blocks on the computers that are connected to the Internet and World Wide Web as a means to prevent student access to inappropriate images and information. Also problematic in World Wide Web research, unfortunately is the issue of credibility. If our students elect to use World Wide Web reference sources to support information or images in their class assignments we require that they use at least two sources for each instance. This is one way we address the issue of credibility and World Wide Web content our students locate and use to support their work. We also require our students to include complete citation information to support their unit and lesson plans as well as their research and artmaking activities. Teachers must also take into account copyright considerations when they decide to use music videos or segments from television programs in the art classroom.[10] Fair use or section 107 of the U.S. copyright law contains a list of the various purposes for which the reproduction of a particular work may be considered “fair,” such as criticism, comment, news reporting, teaching, scholarship, and research.

> The 1961 Report of the Register of Copyrights on the General Revision of the U.S. Copyright Law cites examples of activities that courts have regarded as fair use: “quotation of excerpts in a review or criticism for purposes of illustration or comment; quotation of short

passages in a scholarly or technical work, for illustration or clarification of the author's observations; use in a parody of some of the content of the work parodied; summary of an address or article, with brief quotations, in a news report; reproduction by a library of a portion of a work to replace part of a damaged copy; reproduction by a teacher or student of a small part of a work to illustrate a lesson; reproduction of a work in legislative or judicial proceedings or reports; incidental and fortuitous reproduction, in a newsreel or broadcast, of a work located in the scene of an event being reported" (U.S. Gov. 1999, para. 8).

Teachers should also acknowledge parental and school administrative responses to the use of digital media and visual culture in classroom practice. Open communication with school administration, teachers, parents, and guardians is essential especially when teachers consider anything that may be perceived as controversial. For example, in Taylor's (1999, 2000) study, high school students began their hypertextual approach to art study through an encounter with Madonna's music video *Bedtime Story*. Because of Madonna's controversial personae and the unconventional approach of studying a music video in schools at the time, Taylor first approached the school principal and succinctly explained the unit plan, objectives, links to standards of learning, and rationale for using this in the classroom. She also sent letters home with her students with similar explanations and invited parents and guardians to participate in the process. Announcements and invitations were also shared with school faculty and staff. As a result, the school and community not only knew about what was happening in the art class, but they also participated directly and indirectly through conversations with students and each other. Administrators and faculty came frequently to the art room to see what was happening. In the process, they felt informed about the content and learning in the unit and could talk knowledgably about the interdisciplinary project with others.

In summary, we believe that the promotion of interdisciplinarity through digital computer technology:

- begins and ends with the computer as both a tool and a means for designing connective approaches to inquiry on the part of the teacher and the student.

- is made up of units of instruction that center on a big idea or key concept either evoked from or meaningfully connected to a work of art and linked to both prior and future curricula.

- relies on and links together a range of lessons involving students in historical inquiry, interdisciplinary investigations, artmaking, technique and media exploration.

- requires that students and teachers make and create meaningful connections and links between their art study and other realms of experience both in and outside of school.

- requires that the students and teachers respond to and connect ideas with each other in their computer constructions.

Discussion Questions and Activities

1. Interactive digital computer technology changes every day. Conduct a search online and in current print journals for some of the newest technological tools and software. How could these new tools assist your students in their study of art and visual culture? How would you incorporate these tools in your teaching?

2. On paper, draw a map of one of your units of instruction using shapes to house specific lessons, skills, media, and objectives. Next, create lines to show connections between these shapes. Are there "ways of knowing" in this map that relate to other academic disciplines? Create more shapes that show these areas of study and draw lines between specific areas of the unit where these interdisciplinary connections can be made. Use several pieces of paper if needed and then try and connect the pages

somehow to reveal links to all the academic standards of learning that apply to this unit.

3. Now take the pages that you created in #2 and create a virtual three-dimensional map of this unit using whatever software you feel is appropriate or that you have at hand.

4. Exchange your paper and virtual unit maps with another teacher or pre-service teacher and using another color of pen or pencil (or computer text) make connections between your two units.

5. Find a World Wide Web chat or bulletin board used by teachers and report to the class on the conversations that are taking place. What are teachers most concerned about? What do they need? What do they want?

6. Research the technology staff development opportunities available to teachers in your area. Talk to teachers and again find out what they need and want. Compare their needs and wants to the staff development opportunities available to in your school or district. Design your own staff development technology workshop using art and technology as a catalyst for learning across disciplines.

Notes

[1] For more about how music videos can be used in the art curriculum see Taylor, 2000.

[2] About 21 million teens use the Internet and half of them say they go online every day. 51 percent of online teens live in homes with broadband connections. 81 percent of wired teens play games online, which is 52 percent higher than four years ago. 76 percent of online teens get news online, which is 38 percent higher than four years ago. 43 percent have made purchases online, which is 71 percent higher than four years ago. 31 percent use the Internet to get health information, which is 47 percent higher than four years ago. (Retrieved on-line, November 23, 2005 http://www.pewtrusts.org/) According to a study by the Pew Charitable Trust, "The number of teenagers using the Internet has grown 24 percent in the past four years and 87 percent of those between the ages of 12 and 17 are online. Compared to four years ago, teens' use of the Internet has intensified and broadened as they log on more often and do more things when they are online." This statement comes from a review of the report. To view the complete report, go to http://www.pewinternet.org/report_display.asp?r=162.

[3] See Wass & Cady, 2005 for more information.

[4] Take a look at how photographer Wendy Ewald works with students http://globetrotter.berkeley.edu/Ewald/kids&photo.html.

[5] The World Wide Web on the Internet is composed of documents that have been written in a hypertext markup language (html) that defines everything from links to type style and size. To date basic hypertext applications include Inspiration™, Hypermedia™, Powerpoint™, Keynote™, Storyspace™, Tinderbox™, Weblogs and online learning environments such as WebCt™ and Blackboard™.

[6] For more on the role of experience in learning through the ideas of John Dewey, we recommend, *Experience and Education* (New York: Touchstone, Simon & Schuster, 1938/1997).

[7] Take a look at these texts: Heidi Hayes Jacobs, "Redefining the Map Through Essential Questions," *Mapping the big picture: Integrating curriculum & assessment, K-12* (Alexandria, VA: Association for Supervision & Curriculum Development, 1998) and Stewart & Walker, *Rethinking Curriculum in Art Education* (Worcester, MA: Davis, 2004).

[8] Barrett, T. (1994). Principles for interpreting art. *Art Education, 46*(1), 8-13. Carpenter, B. S. (1996). A meta-critical analysis of ceramics criticism for art education: Toward an interpretive methodology. Unpublished doctoral thesis. The Pennsylvania State University.

[9] Podcasting is the distribution of sound and/or video over the Internet. Its name references Apple's iPod, a phenomenally popular handheld music and video device.

[10] Excellent information regarding copyright issues and citation procedures as well as ideas for incorporating music videos in history classes can be found at http://www.aufdenspring.com/video.html

Chapter 9
INTERDISCIPLINARY LINKING THROUGH VISUAL CULTURE

Nutrition was an important topic in Janice's high school health class. In response to the National Health Education Standard # 4: "Students will analyze the influence of culture, media, technology and other factors on health" (Education World, 2003), Janice's health teacher required her students to read nutrition and ingredient labels on pre-packaged food products. Janice brought in her favorite Kashi GoLean Crunch cereal box. She observed that one cup of cereal contained 190 calories, 25 of those calories came from fat. Each serving contained 3 grams of fat. There were no saturated fats, trans fats, or cholesterol. Each cup of the cereal contained 95 milligrams of sodium, 300 milligrams of potassium and 36 grams of carbohydrates. 9 grams of protein, 8 grams of dietary fiber, 3 grams of soluble fiber, 5 grams of insoluble fiber, and 13 grams of sugar. The values or percentages of daily-recommended vitamins and foods were also listed on the nutrition label of the box. For example, a serving contained 10% of the iron, magnesium and phosphorus, 4% of the calcium, and 8% of the copper recommended for daily nutrition. (See Figure 9.1.).

What are we talking about now? How do cereal boxes correspond with interdisciplinary approaches to teaching high school art? What do they have to do with art? Like the movie references in Chapter 1, the remote-controlled airplanes and gardens in Chapters 2 and 3, and the music videos in Chapters 5 and 8, cereal boxes are a part of our visual culture. Our students know them and see them every day.

Culture — Relevance — Art Criticism

Critical interpretation — Elements and Principles

In art education, visual culture has been labeled as a movement, an approach, a pedagogical stance, and a curriculum. Although it is not easy to define, we can think of the objects, events, images, logos, and experiences in which the visual is central as (1) the idea or concept of visual culture, (2) our visual culture, and (3) a way of teaching with, through, and about visual culture. Said another way, the study of visual culture may involve critical investigations of ideas portrayed though, as well as identification and classification of cultural norms and implications embodied within a visual object, scene, event, or experience. It may suggest a critical exploration of the role, implications, and effects visual culture has on our lives. And/or it may suggest a curriculum through which this content can be taught, interpreted, and learned (Carpenter 2005, Tavin 2003). That is to say that visual culture does not have a single, distinct definition and is concerned as much with objects and visual examples as it is with the social, cultural, political, philosophical, economical, and historical ideas, issues, and concerns related to investigations about visual culture and material.

Understanding and coming to know the social and political implications of our visual culture is important to what Paul Duncum (2002) called Visual Culture Art Education or VCAE. "A visual culture approach requires a substantial shift in what is to be known about images. Knowing about television productions and audience reception is different from knowing about Monet. . . VCAE assumes that visual representations are sites of ideological struggle that can be as deplorable as they can be praiseworthy" (pp. 7-8).

What ideological struggles are there in a cereal box? Art educator and scholar Terry Barrett (2003) used cereal boxes as an impetus for understanding connotations and denotations in art criticism. His students listed what they actually saw (denotation) and then discussed what the elements in the images may have actually meant (connotation). For example, Janice's Kashi GoLean Crunch cereal box featured an enlarged photograph of a bowl filled with cereal, splashing milk, and blueberries as well as text printed in overlapping gold circles on the box that stated Protein 9g, Fiber 8g, and Soy Protein 6.25g. As Barrett predicts, in her assignment Janice saw or associated freshness and health with the image and the text. As the brand name contained the word "lean" and other text on the box stated "naturally sweetened," Janice assumed that this cereal was wholesome, natural, and low-calorie. Therefore, the ideas or ideological implications of the Kashi GoLean Crunch cereal box pertained to health, nutrition, and diet. But, where were the ideological struggles that Duncum (2002) challenged us to find?

Now we catch up with Janice after class. She has little time to go to her locker between classes, so she decides to carry the cereal box with her to all of her classes for the rest of the day. In geography class, Janice's teacher discusses "the changes that occur in the meaning, use, distribution, and importance of resources" (Education World, 2003). Kashi Sales in La Jolla, California distributes Kashi GoLean Crunch cereal. Although not stated on the box, Janice remembers seeing large tractor-trailer trucks featuring the Kashi logo on the highway and therefore assumes that these trucks deliver the cereal directly to stores or to wholesale warehouses near her home.

Janice reports that her grandmother could not find Kashi brands in the grocery store where she shops. Her grandmother lives in a rural area that does not have large grocery store chains and therefore is unable to purchase many of the name-brand items she sees advertised on television. The World Wide Web address featured on the cereal box led Janice to the following mission statement:

Kashi Company was founded in 1984 on the belief that everyone has the power to make healthful changes. At Kashi Company, our mission is to provide great tasting, all natural and innovative foods that enable people to achieve optimal health, wellness and weight management goals. All Kashi products are natural, minimally processed, and free of highly refined sugars, artificial additives and preservatives. (Kashi 2005a, para. 1).

Although orders for Kashi products can be placed through the World Wide Web site, Janice's grandmother does not have a computer. In other words, her geographical location prohibits her from having "the power to make healthful changes" with this particular product. This situation for Janice's grandmother can be considered an ideological struggle.

Another issue comes to light in Janice's economics class where they are studying the role of competition and how "the introduction of new products and production methods by entrepreneurs is an important form of competition and is a source of technological progress and economic growth" (Education World, 2003). On the Kashi World Wide Web site, Janice discovered a new Kashi cereal called Mighty Bites. "With its crunchy fun shapes and delicious honey crunch or cinnamon flavors, Mighty Bites is a tasty way to teach your kids about healthy eating" (Kashi 2005b, para. 2). Referring again to Barrett's approach of looking for connotation and denotation, the image on this cereal features a young boy displaying his flexed arm muscles amidst a cascading stream of milk on which body-shaped cereal figures appeared to be riding. The typestyle is reminiscent of Mighty Mouse or Superman with large bold letters that seem to spring up from the cascading milk. Again, health and nutrition appear to connote the addition of strength, vigor, and fun. The connotation of the images are coupled with the text explaining that not only does this cereal have nutritional value for growing bodies but it:

> . . .contains 35mg of Choline, a nutrient necessary for proper brain development that has been shown to help with memory. Each serving of Mighty Bites provides at least 14% of the daily recommended intake for kids under 8, and the same amount as in 1/2 cup of cooked broccoli, 1/2 cup of cooked brussels sprouts or in 2. oz. of salmon. . . Because of our commitment to providing the healthiest food for your

The study of visual culture may involve looking at the ideas portrayed though as well as identifying or classifying a visual object, scene, event, or experience. It may suggest a critical exploration of the role, implications, and effects visual culture has on our lives.

Photograph by Donna Green, 2006.

> kids, we've partnered with pediatrician, Dr. William Sears, a leading expert in childhood health. We're working together to ensure that our cereal is doing all it can to meet the nutritional needs of your growing child! (Kashi 2005c, para. 2 and 3)

As Janice and her classmates observe, such innovative practice and professional connections relay a sense of urgency to consumers and retailers. The students believe parents must make sure their children eat this mind and body building cereal, and stores must make sure they not only have this product available, but feature it prominently on their shelves. Again, this is an example of an ideological struggle among marketing, commercialization, association, and availability.

Later, Janice takes the cereal box to her health, geography, and economics classes and refers to it when she analyzes the percentages of daily-recommended nutritional values in her afternoon algebra class. In Spanish class, Janice simply translates the text from English, but her Spanish teacher reminds her that a literal translation of the text on the cereal box would not accurately portray the content. For example, “crujido” is a Spanish word that refers to crunch, crack, crackle and creaking typically associated with architecture not food. The Spanish word “cumulo” refers to piles, loads or a grouping together of objects but is not associated with food types such as the Kashi multigrain clusters. Janice's teacher impresses upon her that accurate language translations must be more than exercises of finding words in a dictionary but must also take into account cultural context. (See Chapter 6 for more on cultural context.) When asked to translate the lines “Go Lean Crunch Cereal,” “Multigrain clusters,” and “naturally sweet,” Isabel Cuello, translator in Havana, Cuba responded:

> Literal translation rarely works. If you don't know the context then you can't assure a right translation. Lean has three meanings (inclinarse = to bend magro = low fat. apoyar = to support). On the other hand, go, has also different meanings. Regarding me I would say : Al crujiente y magro cereal! In the second case, it has only one meaning but you have different synonyms, that is, words which even when they have the same meaning, can't be used in the same context because they have different connota-

Photograph by Donna Green, 2006.

> tions. I would say then: Agrupa multiples granos. (email communication, September 23, 2005).

This translation exercise prompts Janice to look at the name of the company "Kashi." The company's web site offers the following explanation:

> In 1984, when Kashi Company founders Philip and Gayle Tauber created their unique Seven Whole Grains and Sesame blend, they considered several names for the product: Gold'n Grains, Kushi, Graino… Finally, they decided on Kashi - a synthesis of "kashruth" or kosher and "Kushi" the last name of the founders of Macrobiotics (a whole foods way of eating/living). Later, they discovered that the word "Kashi" has many meanings around the world. In Russian, the name means porridge. In Hebrew, the word kashruth signifies simple or pure food. In Japanese, Kashi implies energy food. And in Chinese, it means happy food. To us, Kashi means purity and healthy nutrition, and that's just what you'll find in our unique blend of Seven Whole Grains and Sesame. Kashi products are foods for a balanced life. (Kashi, 2005d, para 1 & 2).

Figure 9.1. Janice's cereal box provided a connective learning tool in health, geography, economics, Spanish and art classes. Photograph by Donna Green, 2006.

As she continues her investigation, Janice discovers other meanings of the word Kashi, including a city in China considered the hub of an important commercial district, the western terminus of the main road of the province, and a center for caravan trade with India, Afghanistan, Tajikistan, and Kyrgyzstan. She also discovers the "Mirror of Kashi" or KailÁsanÁtha Sukula's Mirror of KÁD Í, a historical map of VárÁÆasÍ. Printed in 1876 at the press named Vidyodaya, according to some sources the map represents the sacred topography of VárÁÆasÍ with the names and places of deities and water in order to make Kashi constantly visible for people of other locations (Gengnagel, 2001).

In Art, Janice's final class of the day, her cereal box becomes a catalyst for addressing the National Visual Art Standard #2—using knowledge of structures and functions in which:

> Students demonstrate the ability to form and de-

> fend judgments about the characteristics and structures to accomplish commercial, personal, communal, or other purposes of art. Students evaluate the effectiveness of artworks in terms of organizational structures and functions. Students create artworks that use organizational principles and functions to solve specific visual arts problems. Students demonstrate the ability to compare two or more perspectives about the use of organizational principles and functions in artwork and to defend personal evaluations of these perspectives. Students create multiple solutions to specific visual arts problems that demonstrate competence in producing effective relationships between structural choices and artistic functions. (Education World, 2003).

Probably the most fundamental learning connection between Janice's cereal box and the visual art world corresponds to the basic purposes of art. Visual artists create art to be seen. Arthur Danto (1998) proposes that a work of art becomes so only when it becomes subject to interpretation (p. 41). In other words, the fundamental function of a work of art is the portrayal of meaning. That meaning—multiple meanings—can have political, social, and cultural implications and/or be expressions of beauty, explorations of color, and illustrative. Works of visual art function as communicative devices that tell, retell, challenge, suggest, document, express, celebrate, denigrate, beautify, discomfit, revolt, decorate, illustrate, and sell. Artists use whatever tools and methods are available to them—media, technique, elements, principles, exhibition, text, style, symbol, metaphor, and appropriation— to portray these meanings. As Graeme Sullivan (2003) suggests "The artist is a key figure in the creation of new knowledge that has the potential to change the way we see and think" (p. 195).

Artists use whatever tools and methods are available to them—media, technique, elements, principles, exhibition, text, style, symbol, metaphor, and appropriation— to portray meaning.

In art class, Janice and her classmates compare the cereal box designers' methods with those of other artists and/or producers of visual culture. Janice selects *Guernica* by Pablo Picasso. One of her classmates, Mounir, looks at memorial altars on the highway. Janice and Mounir concur with their art teacher's concern that the context and content of *Guernica* and highway altars are very different from the cereal box. They acknowledge, too that such a comparison could be viewed as misleading, erroneous, or trivializing of the important and powerful messages portrayed by Picasso and those who construct memorial altars. However, they argue that all artists and designers have a great deal of power. They point out that an artist's power stems from the way they say something as well as what they express. And they want to analyze how artists and designers provoke anger, mourning, publicity, and even commercial gain.

Janice and *Guernica*. Janice chose Picasso's *Guernica* because she remembered that it was considered to be modern art's most powerful antiwar statement. She also remembered that her art teacher said it was the most important single work of the twentieth century. Picasso painted this mural for the Spanish Pavilion at the 1937 World's Fair. On April 27th, 1937, Hitler's airforce bombed a little Basque village in northern Spain (on behalf of Franco's campaign against the civilian population) and Picasso was incensed. On May 1st, Paris newspapers featured powerful black and white photographs of crumbling buildings and sixteen hundred civilians who were maimed or destroyed in the Nazi bombing practice on Guernica. As a million protesters demonstrated in the streets of Paris, Picasso was provoked to visually capture his outrage at this senseless loss of human life. His large black, white, and gray oil painting (349 x 776 cm or approximately 11 1/2 x 25 1/2 feet) featured stylized, caricaturist, and exaggerated representations of a horse impaled by a javelin, a woman

Figure 9.2. Everyday on his way to school, Mounir passed an altar on the roadside.

wailing over a dead child, and a warrior holding a broken sword as his horse dropped to his knees atop a jumble of body parts. Visual references to a glowing light bulb, extended arm holding a torch, a bull's head, broken bird, and textured surface reminiscent of newsprint text make up the painting that some critics refer to as Picasso's vision of the holocaust at *Guernica* (Stoner, 1999).

Primarily, Janice analyzed and compared the way Picasso and the cereal box designers used the elements of art and principles of design in their works. Designers of the Kashi GoLean Crunch cereal box altered the sans serif typestyle of the word "LEAN"—making each letter thinner than the one before it. This appealed to Janice as she wanted to lose 10 pounds. Janice noticed that the lines in Picasso's *Guernica* were similar in thickness. Although she noticed that he implied some of his lines, she didn't see a connection between Picasso's use of line and the way the painting affected her emotionally. The cereal box designers reshaped the appearance of the cereal box with a brown background behind a silhouetted front label. Because Janice felt that her waist needed to be smaller, she associated this reshaping of the box with a possible change in her own body shape if she ate this cereal. Her exploration of Picasso's use of shape led her to see the ways he reshaped figures and animals to be grotesque and inhuman. The shapes alone did not upset her as much as what they represented. In other words, she knew that they represented the ways that the bombing supposedly reshaped the village, but Picasso went beyond reshaping. She determined that Picasso reduced the people and animals to grotesque symbols and caricatures of pain and suffering. Throughout Janice's process, she found that just talking about line, shape, and other elements of art or principles of design was not enough.

In addition, Janice could link her art experience with the cereal box to ecology (the logo and recycled materials); English and language arts (language used in the description of the cereal); biology (effects of these proteins on human cell function); physics (splashing of the milk, viscosity, boiling temperature, calories and energy), history (comparison of this box with historical advertisements and packaging as well as the changing nutritional directives from the U.S. department of health throughout history); foreign language (compare the adjectives used with meanings and look at the translations used if this cereal box is printed in other languages); and geometry (the organizational structure of the box). If she attended vocational or technical classes Janice could also link her investigation

NOTES:

Visual culture offers teachers a **rich** context in which **inquiry** and **research** based approaches can be developed.

of the cereal to content in business (a closer look at the entrepreneurial approaches and ethical practices of the Kashi company); child care (compare with the nutritional needs of children and look further at Kashi's Mighty Bites); and marketing (look further at advertising). Janice may also have made connections between the information on the food pyramid posters on the cafeteria walls and her cereal box during lunch.

Mounir and Highway Altars. Everyday on his way to school, Mounir passed an altar on the roadside. (See Figure 9.2.). Mounir had seen other roadside altars that consisted of crosses with plastic flowers or more complex structures that included sculptures, items of the deceased, photographs, and candles. The one he passed everyday was more complex. He knew that Sarah created it, a friend who tended to it weekly with the help of the mother of the dead student. A cross, medal, photograph, candles, basketball, gum, candy, plastic flowers, and real flowers symbolized the life and death of Maria, the young teenager who died in a car accident at that location on the highway. Maria's mother visited once a week to keep candles and fresh flowers at the site. After talking with Sarah and Maria's mother about the altar, Mounir researched World Web Wide sites and discovered that some people believe roadside altars distracted drivers and therefore, some states outlaw them. Another young caretaker of the altar told Mounir, "I sometimes wonder if we are doing the right thing by making the altar. My mom thinks we are keeping her spirit here and that she doesn't know she is dead. But I can't let go."

Mounir paid special attention to the medal that hangs on the cross at this highway altar. All last year Mounir himself worked very hard at track practice and in the meets so that he could earn a medal at a meet or a varsity letter for his jacket. He was very interested in the value that Maria's mother obviously shared in such symbols as she prominently displayed this symbol of her daughter's achievement. To Mounir, the "medal" was the focal point or emphasis of the altar and he equated this with the mother's (artist's) desire to celebrate some of the accomplishments of her daughter's short life. Although at this point he began to recognize the value of his teacher's warning to not trivialize highway altars in his comparison with the cereal, Mounir believed the overlapping golden circles on the Kashi GoLean Crunch box (that featured the amounts of protein, fiber, and soy protein in the cereal) were emphasized or featured prominently and because of the color, they looked much like a medal.

Mounir then looked at the way the photograph on the cereal box was cropped on the left side by the silhouetted frame and overlapped the right side. Cereal clusters dropped, floated or moved in a top-right to bottom-left direction. The white splashes of milk framed both the brown organic-shaped cereal pieces and the round, plump, and purple blueberries. The cereal bowl itself was clear glass and gave the viewer an unobstructed view of the cereal and milk. The designers centered the Kashi logo (white serif type on a green rectangle) at the top of the box. The logo contained a simple stylized illustration of a stalk of wheat and a statement that this box was made from recycled cardboard was featured on the label.

Mounir worked to describe the highway altar in a similarly formalist way. He reported that photographs of Maria were scattered throughout the flowers. Some appeared to be torn or battered by rain, wind and other external elements which revealed that the altar is not a static work of art but one that changes daily. In fact, Mounir reported that the altar evoked different, almost contradictory emotions for him depending on the weather. On rainy days, the altar appeared beaten down, lifeless, and mournful, whereas on sunny days the flowers and objects in the altar reflected a celebration, newness, and freshness. Mounir noted also that because of the changeable nature of the altar, he could not categorically identify or describe principles of design, such as placement, direction, or unity. In fact, Mounir concluded that he needed more information about highway and other types of altar art before he could authentically analyze much less compare its formal characteristics. In other words, Mounir could not adequately engage with the highway altar art by merely attending to and/or identifying the artists' use of the principles of design. He needed more.

Ways of Meaning in Art and Visual Culture: Elements and Principles are NOT Enough

What difference does understanding line, shape, color, form, value, texture, and space make when your town has been bombed this morning? It does not. How do emphasis, balance, harmony, variety, movement, rhythm, proportion, and unity relate to a car accident on the highway? They do not. The fact is, "the elements and principles of design were never the universal and timeless descriptors (of art) they were claimed to be" (Gude, 2004, p. 12). As a single approach to art criticism or interpretation, the elements of art and the principles of design do not offer students, critics, art historians, or viewers a robust framework or enough perspective to develop meaningful interpretations and understandings of works of art or visual culture.

This is not to say that the elements and principles of design should be banned from discussions about art and visual culture. Further, we do not suggest that they be replaced by other terms. The fact is that the elements of art and principles of design are not sufficient descriptors or analytical tools for art and/or visual culture. Interdisciplinary contextual analyses require more as do the interdisciplinary curricula upon which such analyses are designed. For example, Olivia Gude (2004) suggests that we add appropriation, juxtaposition, recontextualization, layering, interaction of text and image, hybridity, gazing, and representing to our artists' toolbox. Taylor (2004b) offered the following list of artistic devices associated with technomediation:

> Transparency in media · Erasure · Hypermediacy · Movement through the space · Reperforming and retelling · Varied perspectives and points of view · Repetitive viewing · Inconsistency · Timebased · Repurposing · Rearrange · Instantaneous relocation · Forever changeable · Redefine identity and perspective · Filter · Choice · Multiple views and realities (Multivocality) · Ambiguity · Acceptable untruths · Immersive technology · Networked self (disguised, impatient, entertained, transported) · Endless possibilities · Interrelated and connected · Collaboration ·Representation ·Simulation. (p. 340).

The fact is that the elements of art and principles of design are not sufficient descriptors or analytical tools for art and/or visual culture.

In both cases, Gude (2004) and Taylor (2004b) offer additional lenses through which viewers can examine works of art and visual culture that reveal content that speaks to contemporary concerns that are beyond the perspective of formalist ways of attending to art and visual culture.

In art class, Janice looked at the way the designers of the Kashi GoLean Crunch cereal box used organizational principles and structures of art to portray (1) the nutritional value of the cereal that Janice learned about in health class; (2) the importance of natural resources in geography class; (3) the methods of competition in economics class; and (4) quantitative proportion in algebra class. However, when she tried to do the same thing with the work of art *Guernica*, she could only go so far. Simply looking at the elements of art did not assist or direct her in understanding the contextual and interdisciplinary implications of the masterpiece. Similarly,

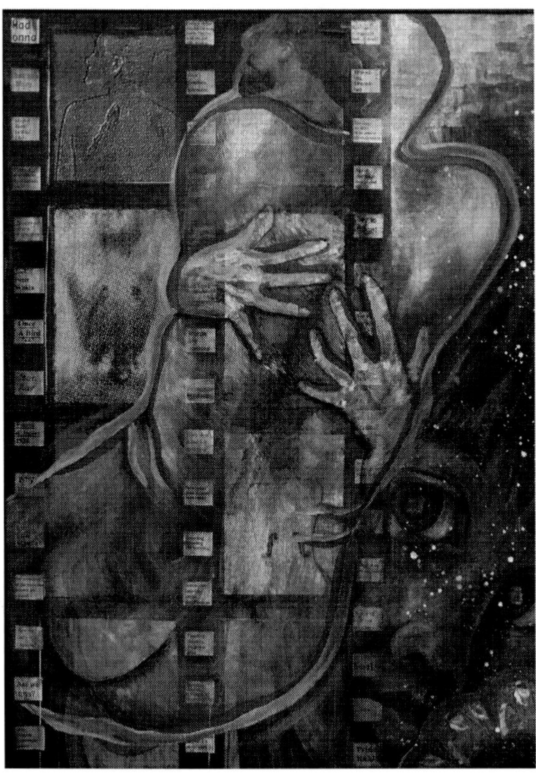

Figure 9.3. Pamela G. Taylor, *Turning Our Backs*, 1999, mixed media. This painting borrows from and appropriates images from works of art and visual culture (Madonna music video) [See Chapter 1 and 8] **to draw attention to and comment upon the self-worth and societal values placed upon women based on reproduction and fertility.**

Mounir's analysis of the organizational principles of art in a highway altar (a work of visual culture) left him frustrated and confused.

In Chapter 5 Interact, we talked about the importance of student-initiated inquiry and collaborative learning. Janice, like many other high school students liked to have a say in what and how she learned. Instead of having to stick with the elements of art when describing, analyzing, and comparing Picasso's *Guernica* with the design of the cereal box, Janice wanted to know more contextual information. In the process, she discovered that the theme of the Paris Exposition where Picasso's painting was featured at the 1937 World's Fair was a celebration of modern technology with an emphasis on the latest advances in aircraft design and engineering. Janice said that Picasso's painting represented an ideological struggle between the promise of technology and its destructive capabilities. Inside the 1937 exhibition that supposedly celebrated technology, Picasso's painting was a testament to its horrific results when placed in the hands of Hitler's air force. Remembering her teacher's warning concerning the comparison of this emotionally-charged work of art with a cereal box, Janice reported that even though her research of the cereal box revealed an ideological struggle (the problem with stating that all people should have access to the best nutrition [supposedly Kashi products] and not making them available in her grandmother's rural area) some ideological struggles require more scrutiny than others.

Similarly, Mounir needed to know more about altar art to adequately respond to the one on a nearby highway. As a result of his research, he discovered that all over the world, temporary altars are often placed at the site of a person's death or where they lived. More familiar to many people were spontaneous public altars such as those at the Oklahoma Bombing site, Columbine High School, John F. Kennedy, Jr's apartment building, NASCAR racetracks honoring Dale Earnhart and Ground Zero of the former World Trade Center Buildings. The construction of spontaneous public altars allows people a place to meet, support, and comfort each other as well as an identifiable site to express their grief.[1]

Mounir reported that permanent altars or monuments are often constructed on specific sites of death and that much dialogue and controversy occurred concerning such plans for the Ground Zero September 11, 2001 site in New York City. Mounir shared that some people believed the site should be treated as a sacred cemetery or burial site whereas others believed the site should somehow illustrate a nation's ability to go on with business-as-usual. For the city officials, civil engineers, developers, survivors, and families of the victims, the negotiation and coalition for healing were essential components in the decision making process of whether the site would remain a grave site, be the location for a new building or other structure, or a little of both. The idea that two inconsistent places could co-exist in the same space was a physical manifestation of ideological struggle that was temporarily, if not permanently, resolved.

According to Kerry Freedman (2003a), education in and through the visual arts is a "process of learning to make and adequately respond to the complexities of the visual arts" (p. xi). And because our contemporary world is filled with visual culture—"it is important for human beings to be able to understand how visual culture images mediate social relationships between and among makers and viewers and among viewers" (Freedman, 2003a, p. 3).

> From this perspective, context is not peripheral to visual culture, or any given work of art; it is a part of visual culture. Contexts provide the conceptual connections that make images and objects worthy of study and are as much a part of a work of art as its form, function, or symbolic meaning. It is a type of narrative loosely attached to the physical object, as is a story portrayed by its image. . .As a result, in new contexts, responses to visual culture are not only formed, they are transformed. [Italics in the original] (Freedman, 2003a, p. 51).

Janice transformed her response to Picasso's *Guernica* when she looked more closely at the exhibition context of the painting. As Janice explored the painting in an interdisciplinary way, she reflected upon her Spanish class activities and began the process of translating images. For example, Janice reported that the Spanish word for lightbulb is bombia. According to art historian Tomas Lloren, "'bombia' is like the diminutive of 'bomb.' So, 'bomba-bombia' is a verbal poetic metaphor for the terrifying power of technology to destroy us" (Lloren in Stoner 1999, para. 11).

Interdisciplinary Ways of Approaching Visual Culture

When studying the environmental art of artists such as Dominique Mazeud *(The Cleansing of the Rio Grande)*, Rebecca Howland *(Environmental Place Setting)*, and Merle Laderman Ukeles *(Methanogenesis)* students in a high school art class also looked closely at such visual culture as magazine advertisements. In particular, they looked at advertisements for scented pink garbage bags. The students worked to create their own environmentally sensitive sculptures in an area outside their classroom (Taylor, 1997). In this process the students studied ecology and researched the decay rate of such materials as leaves, apples, newspaper, and paint. As a result of their research on Ukeles' *Methanogenesis* project the students studied the scientific relevance of one of Allessandro Volta's (1745-1827) experiments in which he discovered and isolated methane gas from the decaying leaves on the bottom of a pond by capturing the bubbles in a device that could be lighted.

Although we consider this example to be an exemplary interdisciplinary art lesson, we believe that the visual culture reference—the scented pink garbage bag advertisement—offered more interdisciplinary possibilities than were utilized. What if the art teacher started with the magazine advertisement and challenged her students to take copies of it to their other classes? What if the students researched and discovered artist Jud Nelson's marble sculpture of a large filled trash bag entitled *Hefty 2-ply*? Perhaps in Spanish or Geography class, students learned how other countries deal with their waste. And perhaps, too they could organize a school-wide cleanup or recycling campaign that would also include participation from community leaders and businesses.

Kathleen F. Ragusea of Sugarloaf Key, Florida, developed a course entitled "Art & War" for The Delta Program in State College, PA. **(Teacher contribution.)** In it, her students examined relationships be-

NOTES:

Works of visual culture include, but are not limited to television, newspapers, billboards, magazines, amusement parks, logos, menus, graffiti, wall murals, memorials, book covers, comic books, cartoons, packaging, tourism, yard art, yard sales, flea markets, bulletin boards, decorated doors, holiday decorations, fashion, set design, movies, film, car art, tatoos, hair, lockers, hallways, floats, homecoming parades, pageants, cemeteries, street fairs, t-shirts, restaurants, shops, chia pets, infomercials, computer and video games, and place settings.

Photograph by Donna Green, 2006

tween art and power in case studies that unfolded across history (e.g. the art of ancient societies, Goya vs. David in the Napoleonic Era, the revolt of the avant-garde during WWI, the totalitarian subversion of art in Hitler's Germany). One way to tie visual culture to this unit might include an exploration of historic and contemporary military recruitment print and media advertisements that pay particular attention to whom these ads were and are directed. Lessons designed to enable students to reveal the ways that the military targeted their ads to parents during the Iraqi war and to young men during the Vietnam War could provide excellent opportunities for the development of critical viewing practices. Teachers could also easily make interdisciplinary connections to history classes. In addition, technical or vocational advertising and marketing classes could benefit from looking at the commercial marketing strategies and devices used in these posters.

Manchester High School West (New Hampshire) art teacher Claudia Michael shared her interdisciplinary unit surrounding Chinese Ancestor portraits:

> I introduce the students to Chinese Ancestor portraits and the tradition of this iconography. We compare and contrast contemporary portraits like those painted by Chuck Close and Alex Katz and Ancestor Portraits. I teach them about Confucian values and how Confucianism creates family and societal hierarchies. From these discussions, the students are also asked to reflect on their own families, American Culture and their own spiritual beliefs (personal email communication, September 17, 2004). **(Teacher contribution.)**

According to art reviewer Margie Romero (2003), Chinese Ancestor Portraits "were part of a religious ritual in a culture that worshipped its ancestors. . .The practice of kowtowing, in which one kneels and touches the forehead to the ground as an expression of submission, was done before the painting because it was believed that the ancestor could bring good luck and wealth" (para.3). Romero suggested that contemporary society kowtows in its own way to celebrities as idols. "Some people among us believe that emulating these idols will help them achieve fame and fortune" (para. 4). Research and discussions about celebrity fandom, particularly fan fiction in visual culture (Jenkins, 1992) could provide an interdisciplinary link between Michael's Chinese Ancestor Portrait unit and English and writing class.

Like Mounir, when Chris studied the art of Ancient Egypt in his high school Art class, he described highway memorials he saw as he and his family traveled to visit an aunt and uncle in another state. Chris searched the World Wide Web and found photographs of highway memorials that featured flowers, photographs, toys such as Teddy bears, and other memorabilia associated with the person who had lost their life on that particular area of the highway. He and the other students compared these memorials to the "Book of Dead" where scenes of the deceased's life were often incorporated into the hieroglyphic scroll that pictured an account of his or her life's deeds before a panel of judges in Ancient Egypt. Other interdisciplinary connections between Chris' study of the highway altars might include links to geography, highway accident statistical analyses in math classes, comparisons to other memorials in world history, and a further study of roadside altars as sites of worship such as those found in Greece as well as iconographic cultural studies in social studies class.

Teachers can help students view examples of visual culture as the focal points and inspirations for units of instruction. For example, in one of our undergraduate pre-service courses, one student, Kristina, constructed a unit of instruction for secondary students based on one of several commercials for Boost Mobile™ cell phones. The commercial Kristina used depicts dozens of senior citizens sitting on and around sports cars, riding customized bicycles, and dancing in what appears to be a street party. Throughout the 30 second commercial, a woman recounts the story of how she recently used her Boost Mobile™ phone to "two-way her friend" as a hip-hop song plays in the background. The final frames of the commercial show another woman dancing "bump and grind style" between two men. As they dance, the text, "Boost Mobile. Designed for young people." is superimposed over their image followed by the phrase "But it's just more fun showing old people."

According to the company,

> Boost Mobile is now recognized as a brand that embraces diverse youth culture, lifestyles and interests. Our knowledge of, and respect from multicultural American youth has generated contagious enthusiasm and rampant success, making Boost the fastest growing wireless provider per capita in the U.S. (Retrieved March 17, 2005 from http://www.boostmobile.com/about_background.html).

This and other Boost Mobile commercials can be viewed on-line (Retrieved March 17, 2005, from http://www.boostmobile.com/commercials.html). Boost Mobile™ has produced other commercials that juxtapose older men and women engaged in youth and hip-hop related activities such as rap concert after parties, skateboarding, and drag racing. Kristina's unit and commercial revolve around the central question, "In what ways are stereotypes and veiled racist characterizations used to promote ideas and ideals to consumers?"

Visual culture offers teachers a rich context in which inquiry and research based approaches can be developed. From teaching in and through music video (See Figure 9.3.) as discussed in Chapter 8 to analyzing a cereal box, as we discussed earlier in this chapter, we have only hinted at the many important aspects and examples of visual culture studies—too many to include in this one chapter. For example, many secondary school students see television as reality (Levin, 2001) and for most it functions as a national curriculum (Freedman, 2004). We recommend that readers refer to many of the visual culture and cultural studies books, chapters, articles and manuscripts available on the Internet and in libraries. Kerry Freedman's (2003a) book *Teaching Visual Culture* is among the first of many visual culture provocations that offer challenges as well as exemplary teaching and learning models. Works of visual culture include, but are not limited to television, newspapers, billboards, magazines, amusement parks, logos, menus, graffiti, wall murals, memorials, book covers, comic books, cartoons, packaging, tourism, yard art, yard sales, flea markets, bulletin boards, decorated doors, holiday decorations, fashion, set design, movies, film, car art, tatoos, hair, lockers, hallways, floats, homecoming parades, pageants, cemeteries, street fairs, t-shirts, restaurants, shops, chia pets, infomercials, computer and video games, and place settings.

In summary, an interdisciplinary approach to teaching art to, from, with, and about visual culture is an exciting method for both teacher and student. Teachers can approach interdisciplinary instruction in this way if they 1) integrate visual culture in an existing unit plan; 2) begin with a work of visual culture to be analyzed; 3) incorporate visual culture that is used in another class; 4) broaden assignments to make space for student initiated inquiry; 5) encourage students to make contextual associations with other works of art(s); and 6) enact evaluation and assessment strategies that we discuss in the next chapter.[2]

Discussion Questions and Activities

1. One of the most vehement objections to the incorporation of visual culture studies in art education concerns issues of aesthetics. What constitutes a work of art? Why do artists make art? Is a Kashi GoLean Crunch cereal box a work of art? When or how can a Kashi GoLean Crunch cereal box become a work of art? If it is not a work of art, why study it in art class? Choose a work of visual culture and ask the same questions.

2. Look at one of your unit plans and incorporate a work of visual culture. Why did you choose this particular work of visual culture? How does it relate to your original unit plan? How does the visual culture change or inform the content or rationale of your unit? What are the ideological struggles inherent in the visual culture? How does it assist your students in understanding social and political implications of the art world? Of their world?

3. Now look at a work of visual culture and formulate a plan for using it as the center of an interdisciplinary art unit. As you look at varied works of visual culture, talk with a teacher or student in another discipline and share the different links you see between your respective subjects. How does it relate to the standards of unit in your class? How could it relate to the standards in another class?

4. Select a work of art from an art history text or art survey book. Look through magazines and find an image that seems to relate to that work of art in terms of their formal characteristics or subject matter. What conceptual and content connections can you make between the work of art and the image? In what ways do they "mean" in similar ways? Use these two images as the starting point for an interdisciplinary lesson or unit of instruction.

Notes

[1]See Ballengee-Morris (in press), "Exploring Everyday Altars" in the *Journal of Cultural Studies*.

[2] In recent years many articles have been published about visual culture in art education. In particular, we recommend entire journal issues devoted to the subject such as *Studies in Art Education*, 44(3) and *Art Education*, 56(2) and 58(6). Among the numerous journal articles on visual culture in art education include Ballengee-Morris & Stuhr, 2001; Barrett, 2003; Bolin, 2003; Brown, 2003; Chapman, 2003; Duncum, 1999, 2001a, 2002b, 2002, 2003, 2004; Freedman, 1999, 2000, 2003a, 2003b; Heise, 2004; Keifer-Boyd, 2003; Kindler, 2003; Krug, 2003; Parks, 2004; Pauly, 2003; Smith-Shank, 2003; Sullivan 2003; Tavin & Anderson, 2003: Tavin, 2003b, 2004; Taylor, 2000; Taylor & Ballengee-Morris, 2003; Villeneuve, 2003; & Wilson, 2003. And as a basic foundation text on visual culture, try Nicholas Mirzoeff (1999). *An Introduction to Visual Culture*.

Chapter 10

STARTING WITH THE END: STANDARDS, ASSESSMENT AND EVALUATION

A large door suspended above simple white steps greeted visitors to the "Sharing a Place called Home" spring 2000 exhibition at the Radford University Bondurant Center for the Arts gallery in Radford, Virginia. A line of embellished shoes hung on the wall beside a white bed colored by a pieced art quilt. Another quilt displayed dozens of text-patterned papers and Polaroid™ transfers placed in a "friendship" quilt pattern on the wall. Painted window blinds hung on another wall crimped in places to reveal mono-prints of home interiors. A wire cage sat on the floor and contained animals shaped from bread dough. Petitioned shadow boxes grouped together held small objects, representing room designs of such world cultures and nationalities as Cuba, American Indian, Nigeria, and Mexico to create a multicultural village of sorts. Hand-painted chairs dotted the walls and clay chair shapes flanked the steps. Cardboard covered one section of the gallery that contained empty frames with signs and symbols of homelessness. Simple proportioned face designs painted on tiles lined the floor beneath the frames, which caused viewers to step on them to see the frames. Large stamped pillows spotted the floor in front of a large television that looped a video of lesson plans, images, and a low hum of student voices working in the art room.

Text panels hung below, beside or underneath each area of the exhibition installation and asked: What constitutes a home? Who lives there? Who doesn't? Is a house always a home? What is "home away from home?" What is

Objectives — Goals — Flexible Purposing

Formative and Summative — Curriculum

homelessness? Should everyone have a home? How are homes shared? How is the idea of home shared? How can we share our place called home? In addition, other text panels in the exhibition explained the processes involved in each activity, the Virginia Visual Arts Standards of Learning, the links made to Standards of Learning in other academic disciplines, and student quotes about the project. For example, the text panel beside the multicultural village boxes read:

> Students researched and discussed the ways that rituals and the special objects associated with those rituals make a place feel like home. They compared the ways that cultures other than their own have rituals that are similar and different because of location, climate, and cultural values. From this information, they constructed these shadow-box rooms and filled them with personally-created symbolic objects. As one student explained, "I researched Tibetan Nomadic culture and discovered that they wander through such high elevations because there is productive grazing land for their yaks, sheep, goats and horses. Tibetan Nomads take very good care of their herds, and are on-the-move so much that they do not have very many personal posessions to go in their tents. They are known for creating beautiful woven clothing and blankets using wool and natural dyes. Therefore, I created a sparse and airy room and placed a small blanket that I created using an inkle loom." Standards of Learning addressed in this activity include: (Geography) Understand how culture and experience influence people's perceptions of places and regions; (Visual Art) Students analyze common characteristics of visual arts evident across time and among cultural/ethnic groups to formulate analyses, evaluations, and interpretations of meaning; (Foreign Language) Students demonstrate understanding of the concept of culture through comparisons of the cultures studied and their own.

This exhibition of student art was created as a result of an interdisciplinary unit of instruction entitled "Sharing a Place Called Home." It celebrated and showcased students' art and writing. A formal reception provided a forum for parents, teachers, family, and friends to honor their young people and their teachers. Announcements and press releases resulted in local newspaper coverage and served as advocacy for the school's art program. Administrators used the exhibition as an opportunity to meet, mingle, and garner further support for their school and school district. This exhibition was a form of assessment.

Making Learning Visible

According to Campbell and Harris (2001) "The purpose of assessment in thematic instruction is to make learning visible so that schools, teachers, and students can demonstrate student attainment of deep understanding" (p. 155). Indeed, assessment can be thought of as a means to make learning visible. That is, through assessment, students, teachers, administrators, parents, and concerned community members see evidence of what students have learned. In the case of the "Sharing a Place Called Home" unit, assessment took the form of an exhibition.

David Burton, in his book *Exhibiting Student Art* (2006), called exhibition the epitome of authentic assessment because it is a deliberate expression of learning that can be experienced and understood by others. Assessment can and should take many forms. We advocate that teachers explore and employ a variety of assessment tools as they design and implement interdisciplinary curriculum.

We began this chapter with the description of the "Sharing a Place Called Home" exhibition because the formulation of the unit of instruction began with a vision of the exhibition. (See Figures 10.1 and 10.2.). In other words, the beginning of the unit started with the end.

> I had visions of the students' art becoming an installation of a place called home in the gallery. I began looking for sets or props at the outset and imagined the ways that student art could both represent different areas of a home as well as expand upon and connect the ways those areas are meaningful in multiple contexts. In some cases the props served as frames or display areas such as the window blinds, the bed, television, and pet cage. In other areas, the art became the area of the home such as the line of shoes representing a closet, the chairs, and

the nightlights. And in some areas, such as the cardboard cubicle, the interior space was challenged to be anything but a home. When I had a fairly clear vision of the ways these objects or areas could become an installation, I began looking at the standards of learning in the visual arts as well as standards from many of my students' other academic classes. From those I worked to formulate questions, discussions, research, and inquiry-based activities that would involve the students in applying those understandings or ways of knowing to art objects for the installation. For example, they practiced painting techniques when choosing symbols to apply to their shoe art. They researched and shared information about homes in cultures other than their own while exploring relief sculpture techniques when creating the multicultural village. They reviewed mathematical and figural proportion rules as well as ceramic glazing procedures when making the tiles. Along the way, opportunities, problems, and student interests changed many of my preconceived ideas and actually made the experience and the exhibition more meaningful than I envisioned in my original idea (Taylor, personal communication, September 30, 2005).

Although one goal of assessment is to organize learning so that it is visible and can be documented, it is also important to note, that in this assessment-based process, Dewey's (1938) "flexible purposing" or the value of recognizing new and emergent possibilities for learning and knowing is intrinsic to art education. Jacobs (1989) echoes this point. "In my experience, school districts that use a combination of design options manifest the greatest success and the least fragmentation in their programs. Interdisciplinary designs should not be an all-or-nothing proposition"

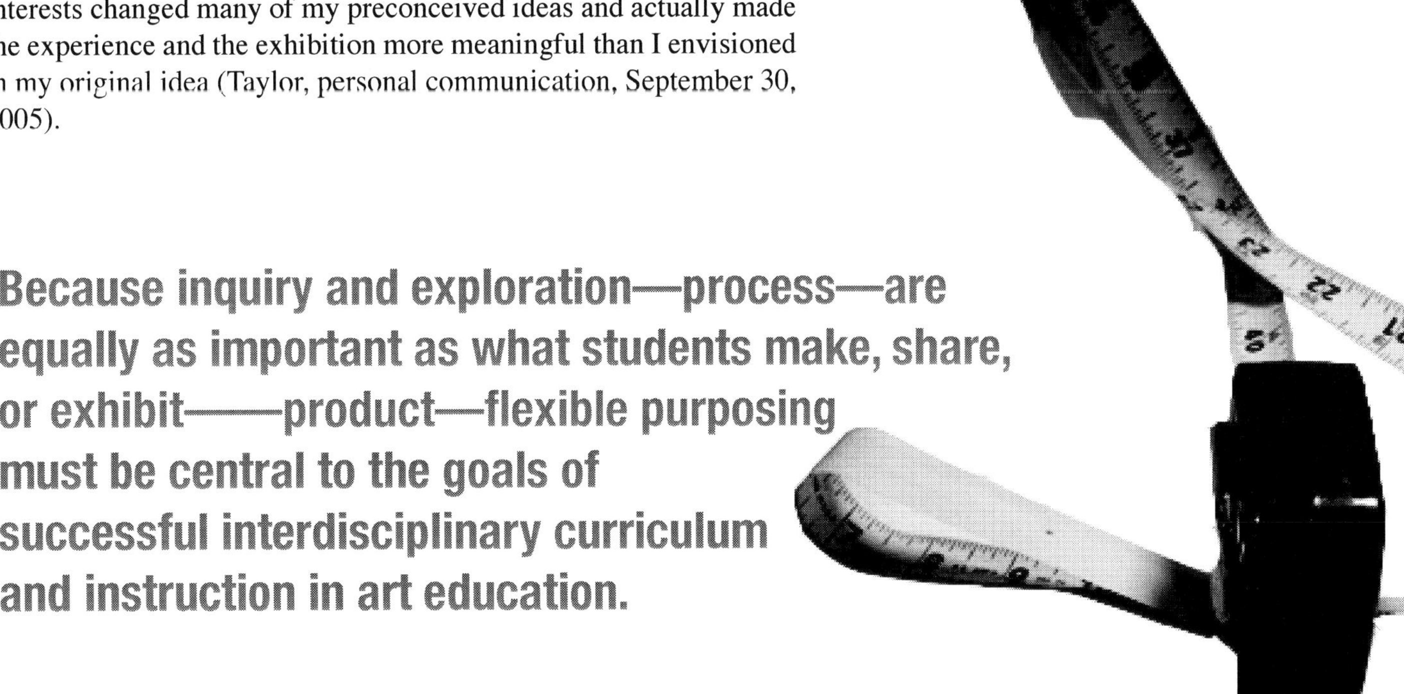

Photograph by Donna Green, 2006.

(Jacobs, 1989, p. 19). And Eisner (2004) cautions, "The kind of thinking that *flexible purposing* requires thrives best in an environment in which the rigid adherence to a plan is not a necessity. As experienced teachers well know, the surest road to hell in a classroom is to stick to the lesson plan no matter what" (p. 6). We hope that the examples in the previous chapters offer a broad picture of the flexibility inherent in an arts based interdisciplinary curriculum.

Granted, knowing generally what art students will create as a result of a lesson or unit is a typical way of approaching curriculum. The difference in the "Sharing a Place Called Home" unit was that the thematic exhibition installation was the goal for the unit and the works of art were not as clearly defined as was the vision of the exhibition.

Clarifying the goals for what students should know and be able to do upon the completion of a unit of instruction is essential to authentic and successful assessment of interdisciplinary approaches. Curriculum theorists Grant Wiggins and Jay McTighe (1998) in their book *Understanding by Design* refer to this approach as backward curriculum design where the curriculum designers identify desired results, determine acceptable evidence, and plan learning experiences and instruction (McTighe & Wiggins 2004, p. 38). And, in Lynn Brown and Julie Craven's (1999) chapter "Beginning from the End" in *Art Works! Interdisciplinary Learning Powered by the Arts* (Wolf and Balick, 1999), students were asked, "What would you need to know in order to meet the design challenge we've given you?" (p. 21). In other words, what should students know and be able to do and how do we get them there? What evidence are we willing to accept as proof of what they know and can do? And to what degree? What kinds of activities, lessons, or exercises, are needed for students to demonstrate what they understand and apply their understanding in an assessable way?

The overarching goal for the "Sharing a Place Called Home" unit was the creation of an installation exhibition that provoked viewers to critically reflect on many of the ideas

Figure 10.1. The goal for the "Sharing a Place Called Home" unit was the creation of an installation exhibition that provoked viewers to critically reflect on many of the ideas that the young artists grappled with in and through the creation of the objects on display.

that the young artists grappled with in and through the creation of the objects on display. It was not simply an exhibition of student artwork. It was a continuation of the learning and understanding process. Wiggins and McTighe (1998) described the six facets of understanding as: can explain; can interpret; can apply; has perspective; can empathize; and has self-knowledge (p. 44). We recommend that curriculum designers consider these six facets when they construct assessment tools for interdisciplinary units of instruction.

Figure 10.2. Cardboard covered one section of the gallery that contained empty frames with signs and symbols of homelessness. Simple proportioned face designs painted on tiles lined the floor beneath the frames, which caused viewers to step on them to read the text.

Because inquiry and exploration—process—are equally as important as what students make, share, or exhibit—product—flexible purposing must be central to the goals of successful interdisciplinary curriculum and instruction in art education. Although developing flexible yet assessable goals for interdisciplinary instruction may appear to be interminable, authentic attempts to do so are essential.

A flexible purpose is a purpose just the same. Every lesson of every unit must have a purpose and goal. Indeed, asking oneself, "Is this worth knowing and why?" should be an essential qualifier for everything we do in education. A flexible assessment-based approach to interdisciplinary high school art teaching involves teachers and students from multiple disciplines to formulate, reformulate, and reflect upon the attainment or effect of important goals and aims for each learning experience.

But making visible what students learn is only one way to consider assessment of interdisciplinary instruction. We can also assess—make visible—what teachers learn or do during instruction, the interdisciplinary curriculum tasks designed and taught by individual and teams of teachers, and the impact of interdisciplinary instruction on the overall school program.

Before we explore various approaches to assessment, let us take a closer look at three important issues that arise whenever we engage our university pre-service teachers and veteran teachers in conversation about assessment. One issue is the current concern for local, state, and national standards. Many teachers are confused about which standards to address as they plan their curricula and are unsure as to how these various standards relate to and with each other. A second issue of importance is the confusion between the terms assessment and evaluation. While these terms are often used interchangeably, understanding the difference between the two is important and necessary in the development of successful interdisciplinary curriculum. A third issue is the perception that many teachers have about assessment. Our experience tells us that many teachers either do not understand assessment and have therefore developed a fear of it or simply have not been taught how and why assessment is important. By taking a closer look at both issues we hope to establish a foundation for looking more closely at a variety of assessment approaches for the high school art.

Assessment is the process or evidence of gathering information about what students have done, made, said, or created. Evaluation is a process or evidence of placing a value on and judging what students have done, made, said, or created. Simply, assessment explains what has been done and evaluation describes the degree to which it was accomplished.

Curriculum Standards

Education in the United States of America is primarily the responsibility of state and local governments.

> It is States and communities, as well as public and private organizations of all kinds, that establish schools and colleges, develop curricula, and determine requirements for enrollment and graduation. The structure of education finance in America reflects this predominant State and local role. Of an estimated $909 billion being spent nationwide on education at all levels for school year 2004-2005, about 90 percent comes from State, local, and private sources. (U.S. Department of Education, 2005, para. 1).

The U.S. Department of Education began in 1867 as a means for helping states establish effective school systems by collecting and disseminating information and data. The then-named Office of Education became responsible for administering support for land-grant colleges and universities as a result of the Second Morrill Act of 1890. The Smith-Hughes Act of 1917 focused on vocational education as the next major area of Federal aid to schools. Agricultural, industrial, and home economics training for high school students became a focus of the 1946 George-Barden Act. Federal support for education significantly increased as a result of World War II. Communities greatly affected by the war received aid according to the Lanham Act in 1941 and the Impact Aid laws of 1950. The "GI Bill" authorized postsecondary education assistance for military veterans in 1944. In response to the Soviet launch of Sputnik, the National Defense Education Act (1958) supported the development of highly trained individuals to help America compete with the Soviet Union in scientific and technical fields through loans to college students, improvement of science, mathematics, and foreign language instruction in elementary and secondary schools, graduate fellowships, foreign language and area studies, and vocational-technical training. In the 1960s and 1970s, anti-poverty and civil rights laws challenged equal access to education for all U.S. citizens.

The anti-poverty and civil rights laws of the 1960s and 1970s brought about a dramatic emergence of the Department's equal access mission. The Elementary and Secondary Education Act of 1965 launched the Title I program of Federal aid to disadvantaged children. The Higher Education Act approved financial aid programs for needy college students. The Department of Education was established as a Cabinet level agency in 1980. The mission of the Department of Education is to ensure equal access to education and to promote educational excellence throughout the nation. They do this by continuing to obtain and disseminate information and data regarding school effectiveness, research, and student performance as well as through the administration of educational programs, grants, and fellowships. (U.S. Department of Education, 2005).

In 1997, The Seven Priorities of the U.S. Department of Education included priority #4, which stated, "All states and their schools will have challenging and clear standards of achievement and accountability for all children and effective strategies for reaching those standards" (U.S. Department of Education, 1997, para 1). In response to this priority, national education organizations and state departments of education implemented sets of standards or guidelines for all academic areas. National standards are typically general and designed to offer states and local school districts guidelines for what students should generally know and be able to do at varied levels in their education. State departments of education then take the national standards and tailor them to state specific guidelines. Local school districts look to the national and state standards to devise objectives and curriculum adding specific local and/or community sets of knowledge and skills. Teachers then look to the curriculum guidelines and organize units and lesson plans that engage students in activities to introduce, provide opportunities for guided practice, and ultimately demonstrate an understanding and proficiency of the standards of learning. In addition, teachers and schools may have their own sets of standards that relate to such areas as the school climate, safety, management, community relationships, behavior, and professionalism.

Clarifying Evaluation and Assessment

Simply put, evaluation and assessment are not the same. Assessment is the process or evidence of gathering information about what students have done, made, said, or created. Evaluation is a process or evidence of placing a value on and judging what students have done, made, said, or created. Simply, assessment explains what has been done and evaluation describes the degree to which it was accomplished. For example, when you say "that was good ice cream" you are assessing what you tasted. When you say, "that was the best ice cream I have ever tasted," you place a value on that particular dessert in relationship to others you have tasted. When we look at student work or the curricula we have implemented we can assess them based on information we gather about them in relationship to our predetermined goals. To evaluate student work and curricula then means to judge and assign a value to them in terms of the degree to which they have met such goals or expectations.

Teachers and Assessment

In our travels to schools and other educational sites we have come to realize that many teachers do not like assessment but have not necessarily explored why they feel this way. For example, one summer during a professional development institute in which one of us was teaching, the participating teachers were asked to work in teams to produce interdisciplinary units of instruction. The curriculum plan was to take the form of a blueprint or floor plan for a house as a metaphorical representation of curriculum. That is, each room or area of the house was to symbolically stand for a different part of the unit. Some units used the refrigerator as the gallery or exhibition space where students would display their work. Others used the kitchen as the location in which the students would "cook up" artworks. For others, the yard served as the metaphorical location in which teachers and students would take

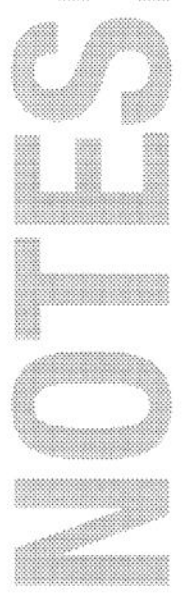

Clarifying the purposes of assessment is fundamental for both teachers and students. Teachers and students are entitled to know the justification for required assessment practices.

field trips. What struck us, and the teachers, is that most of the groups used the bathroom—and the toilet specifically—to depict assessment. Some of the teachers stated that when they hear the word "assessment" they want to "flush it down the toilet." For us, and other teachers, the toilet symbolized a place and means to do away with part of the curriculum we would rather not face or present to the public. Symbolized in this way, assessment is a private matter that is better left behind closed doors. But why is assessment thought of in this way?

"Assessments can be proactive tools of reform, pressing toward new forms of teaching and learning" (Resnick, et. al., 1992, p. 186). Is this what scares us? The fact that assessment may reveal more of what we need to know and do as teachers? Understanding and outlining basic purposes of assessment may assist us in lessening this concern. According to Clarke and Agne (1997), possible purposes of assessment include: "(1) to help students understand what they have learned and what they need to learn; (2) to let parents know how their children are doing; (3) to help teachers compare achievement among students; (4) to establish a basis for tracking; (5) to measure teacher performance; and (6) to set budget priorities" (p. 295). Clarke and Agne reported that the northeast Supervisory District of Vermont outlined nine purposes of assessment within four principles related to individual students and public understanding: "(1) to focus student learning (to guide students effort and to inform and guide parents); (2) to focus teaching (to inform day-to-day teaching and to evaluate teaching effects); (3) to improve supporting systems (to determine special services and evaluate systems and curriculum); (4) to influence policy and planning (to evaluate programs and inform the public)" (Clarke & Agne, 1997, p. 296). An analysis of each purpose included looking at the assessment tools already in place; identifying the function of the tools and the recipients of results; linking these purposes with school goals and objectives; recognizing the strengths and weaknesses of the process; and formulating new ideas or assessment possibilities (p. 297).

We believe that clarifying the purposes of assessment is fundamental for both teachers and students. Like Jeremy's straightforward question in Chapter 2, "Why do I need to learn/know this?" teachers and students are entitled to know the justification for required assessment practices. In other words, "What does this assessment process, activity, or test have to do with my learning, knowing, teaching, and understanding?'

Student Assessment/Instruction

Student assessment "is a direct assessment of the knowledge constructed by students about selected themes in individual classes. When assessing students' gains, teachers always need to emphasize what was learned rather than what was taught" (Campbell and Harris, 2001, p. 155). Among the strongest examples of student assessment we observe are those in which assessment tasks looks like instruction.

Performance Assessment is a form of testing that requires students and/or teachers to perform tasks rather than select pre-determined answers from a test. Many teachers may remember being assessed during their student teaching experience by both cooperating teachers and university supervisors who wrote copious notes throughout the class period. Performance assessment takes many forms however, and includes gauging how students perform specific exercises in the class, how they pay attention to the goals and objectives of the lesson, and how they use the techniques, skills and ways of working introduced to them in class to convey or express meaning in the art they create. Performance, also known as alternative and authentic assessment methodologies encourage a means for examination of student performance on significant tasks that are relevant to life in and outside of school (Worthen, 1993).

Shari Williams teaches art at Benton High School in Benton, Arkansas. She and several of her colleagues developed what they call rich tasks as alternative assessment forms in arts education (art, drama, music, and dance). These tasks include reflective writing, research, technology, group interaction, peer review, and self-evaluation. The tasks are rich with assessment methodologies that are catalysts for interdisciplinary discovery and connection. **[Teacher contribution.]** Performance (or the act of doing) requires multiple ways of knowing that extend beyond the art classroom. Effective assessment is not only the culmination of instruction, but also an important means for discovering relevant and meaningful information by both student and teacher (Armstrong, 1994).

Formative Assessment. As many teachers know, there is more to understanding students' progress than simply looking at or assessing their final or best work. As student/teacher reflection is a valuable dimension of the assessment process, teachers and students might prefer a process-folio or portfolio that documents a student's learning process rather than a "best works" portfolio. Portfolios that contain varied student activity projects—both successes and failures—may serve to demonstrate students' understandings and active participation in the learning process. A nationally recognized example of the process-folio assessment technique began with Harvard's Project Zero and specifically the Arts PROPEL approach (Simmons & Winner, 1992). To begin, students collect more than a diverse body of finished artwork. They assemble biographies of works, a range of works, and reflections. A biography of a work reveals the process of the production of any major project. Reflections are documents or audio or videotapes that result when teachers ask students to look at their collections of work and take the stance of an informed critic or autobiographer. In an art process-folio, the student may bring together journal entries, letters, poems, and essays from language arts or social studies classes. Students note what is characteristic, what has changed with time, or what still remains to be done. At the end of any semester or year, teachers continue to challenge students to take time to study their collections and select several works that best illustrate what changed for the student during that time. Further interviews allow students time to explain their choices. These works, along with student and teacher commentaries, come together as varied forms of data and become a final portfolio that can be passed along as a continuing document from year to year (Simmons & Winner, 1992).

There are many purposes related to the process-folio form of assessment. We identify at least five purposes: (1) providing evidence of growth and learning; (2) offering a place to give and receive feedback; (3) encouraging critical self-reflection and evaluation; (4) documenting depth of thinking involved in the learning; (5) affording opportunities to witness interdisciplinary connections. We recommend that when teachers choose a process-folio form of assessment, they do so with one or more of these purposes in mind.

Because cognitive psychologists believe that learning is an active mental process, an authentic assessment should include a means for knowing and understanding the kind of thinking that is taking place in the classroom. Problem solving advocates believe that the correct assessment should

Because cognitive psychologists believe that learning is an active mental process, an authentic assessment should include a means for knowing and understanding the kind of thinking that is taking place in the classroom.

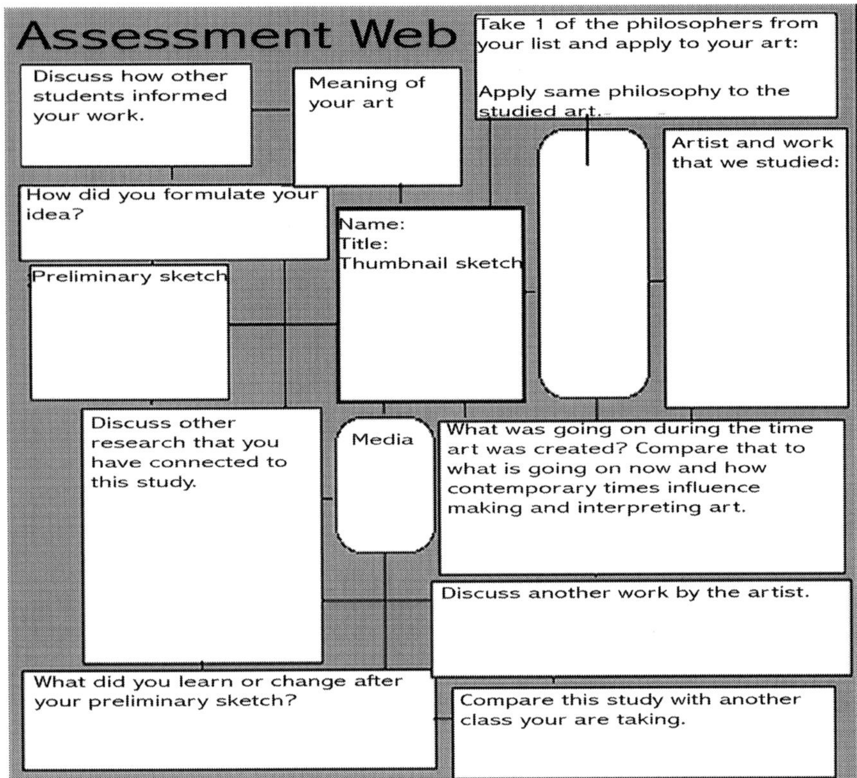

Figure 10.3. Author Pamela G. Taylor's former high school students completed assessment webs following their completion of every unit. These webs hung alongside the students' art both in the classroom and outside the classroom providing a means for viewers as well as evaluators to understand the students' learning processes.

involve a kind of interview or discussion as in the Arts Propel[1] approach, rather than simple observation. They believe that the thinking process simply cannot be understood any other way (Peterson & Swing, 1982).

Students of Linda Hinson, art teacher at Blacksburg (Virginia) High School and Middle School, update their journals every day.[2] They take notes, attach completed handouts and media exercises (value scales and color wheels), write criticism, research, activity instructions, and a signed contract regarding classroom rules in 8" x 10" journals provided by the class. Hinson reviews these journals weekly. She uses a checklist of criteria glued inside the journal and her short but informative individual comments. Hinson involves her students in self-assessment and self-criticism journal activities as well by asking specific questions that relate the activity to the goals and standards outlined in her original instructions and daily reviews. Questions related to interdisciplinary issues are co-formulated with other teachers in the school and/or reflected through directions that ask students to make connections between specific art activities and other classes and/or realms of experience.

Teachers make observations of what their students do every day. In the art classroom, teachers spend a great deal of time walking around the room helping, explaining directions, reiterating important points, asking questions, and offering advice. As they do so, teachers assess what is working and what is not working and may stop the class to explain specific issues. Opportunities may also be discovered for taking the unit or lesson beyond the original goals, especially when they relate to interdisciplinary study. For example, in an activity where students were exploring their multiple forms of identity (ie: son, daughter, friend, jokester, cheerleader, rock star, chess player) through the creation of multiple self-portraits, the teacher noticed that many students struggled with using rulers to divide their paper in equal rectangles. Upon closer observation, the teacher realized that none of the tenth graders in the class knew how to read a ruler and that they were very embarrassed by this fact. In addition to helping these students individually, the teacher shared this information with the math teacher. Upon further investigation, the math teacher discovered that this particular class missed this important instruction as the result of a curricular restructuring process several years prior and no one had noticed this issue earlier. As a result of this art teacher's observation and assessment, all tenth grade math classes incorporated this important skill immediately in their curriculum that same year.

In Chapter 5, we talked about the importance of varying instructional strategies in the interdisciplinary high school art classroom. In schools that implement longer blocks of instructional time, varying instructional assessment strategies that involve observation are productive for the teacher and for student progress and redirection. For example, in-process critiques in the middle of a class period or any time during an activity that lasts several days provides a forum for students to see the ways that other students interpret directions and share issues, concerns, and ideas that inform each other while at the same time let the teacher know what is working and what is not working in the unit or lesson. During such critiques, interdisciplinary opportunities may often arise as both a means for further expression of the students and a way for teachers to clarify directions and goals. When these in-process critiques are directly linked with the criteria in which the activity will be graded, students may alter or redirect their work in time to affect their future grade.

Summative Assessment. Teachers usually assign students their course grades based on some form of summative assessment. Typically, there are many grades recorded for activities, handouts, exercises, works of art, papers, quizzes, and tests that are averaged. Grading of individual assignments is typically based on criteria, directions, and expectations.

Carole Henry (1990, 1992) reported on the use of criteria checklists in which a student's performance in varied aspects of each assignment received a certain number of points. Similar to rubrics, such criteria checklists assist students in knowing exactly what is expected of them. They also expedite the grading process for teachers who simply look for evidence and mark according to the degree to which the student demonstrated their understanding of specific criteria. For instance, when grading the student personal perspective works of art [See Chapter 1], the criteria sheet may include (1) one-point perspective; (2) at least three other methods of perspective (overlapping, color, placement, value, size, etc.); (3) figural proportion; and (4) artisanship. But, the unit of instruction involved the students in much more than what can be outlined in these criteria. "Works of art are not the whole of art; they are only its rare peaks" (Arnheim, 1969, p. 295). As the goals of the unit included understanding and making connections among interdisciplinary, art, and personal ideas and values, an examination of only the final work of art does not adequately address the purpose of this assessment. Perhaps, in addition to submitting the work of art, students could also complete a handout with specific questions in which they explain the ways they, for example represented how their perspective was altered or influenced by the people they chose to represent in their drawing. Criteria for such open-ended questions can be established (Henry, 1990, 1992) and may include: (1) links between personal ideas and those of the figures the student chose to represent; (2) evidence of ideas being informed by research; (3) grammar, spelling, and punctuation; (4) explanation of mathematical perspective devices used in the art work; (5) ability to explain their choice of media and technique in expression.

Using a similar idea to prepare her high school art students to use computer hypertext [See Chapter 8], Taylor (1999) created a generic assessment web on paper. (See Figure 10.3.). Created with questions, instructions, and blank spaces for responses connected by lines and arrows, each student completed and submitted the required web with their final projects. The thirteen boxes asked the student to: create a thumbnail sketch, your name and title of the work; take one of the philosophers from our list of aesthetic stances and apply that theory of art to your work; apply that same philosophy to the work(s) of art we studied in class; describe the artist and work of art that we studied in class; what was going on during the time the art was created and how does that compare to what is going on now and how your work

NOTES:

Photograph by Donna Green, 2006

Assessment is more than a test or quiz that comes at the end of a unit. At its very best, assessment should be directly tied to the objectives and goals of an activity, lesson, unit or curriculum plan. And without a doubt, assessment should be part of the learning experience for students.

was influenced?; discuss another work by the artist and explain how it informed your work; compare the work and study we are doing in the art class with at least one other class; describe and justify your choice of media; create a thumbnail of your preliminary sketch; what did you learn or change after your preliminary sketch?; discuss other research that you have connected with this study; what is the meaning of your work?; how did you formulate your idea?; discuss how the other students in the class informed your work (Taylor 1999, p. 105).

For assessment purposes, teachers and administrators may rely on student portfolios that reveal the body of work created during the class as well as data from questionnaires and surveys such as those used at the beginning of the course to ascertain student interest, knowledge base, and skill level.

Standardized testing such as the current SOL (Standards of Learning) tests in many states have, to date, contained little if any art skill or content assessment measures. Many school districts, states, and national organizations are currently working to develop arts-based standardized testing assessments.

The 1997 NAEP (National Association of Educational Progress) arts education framework followed the National Visual Arts Standards in the development of the content and processes to be assessed. Evaluative exercises were developed to demonstrate a student's ability to apply his or her knowledge of art contexts (personal, social, cultural, historical), aesthetics, form and structure, and processes to their cognitive, affective, and motor skills (perceptual, intellectual/reflective, expressive and technical) (Council of Chief State School Officers, et al., 1994, p. 30). In other words, the goal of the NAEP performance assessment was to enable the evaluator to understand how students translate their contextual learning in the art class to their skill and technique of making art. "Performance exercises should be rigorous, demanding, authentic, and should require students to apply and demonstrate what they know and are able to do" (p. 35). The NAEP categorized these exercises into (1) production, (2) questions requiring short answers, and (3) questions requiring extended written answers. An authentic assessment would involve a student experiencing a combination of formats.

One example of a visual arts exercise to understand a student's ability to develop ideas for personal imaginative symbols involved a minimal script relating the ideas of how symbols were used in a Renaissance work, Cardinal Albrecht as St. Jerome, completed in 1526 by Lucas Cranach the Elder. Another and more contemporary work, a photograph of comedian Steve Martin by Annie Liebowitz, was also explained minimally with regards to the use of symbols. The students were then instructed to create three 1 1/2" by 2" drawings of symbols that expressed something about their interests or personality. Another more finished drawing (3" x 6") was also required with directions to use one or more symbols in a composition that expressed something about their interests or personality. Two questions asked why certain symbols were chosen and how the technique, style and colors they used helped the symbols communicate something about them. Each of these activities was timed and evaluation criteria included levels of proficiency from not developing the sketches to providing excellent evidence of choice and rational for technique and color. Other examples of NAEP exercises included: illustrating a story; expressive use of line; color used to express mood and feeling; double self-portraits depicting two contrasting feelings; a design problem involving the use of chairs to create space for a variety of social situations; locating presented art in a timeline; a more involved explanation of style indicative of historical period; comparisons of works critically and aesthetically; and an exercise involving form and function.

The purposes of this national assessment could be considered two-fold. A secondary analysis of the 1997 NAEP Arts assessment (Diket, Sabol, Burton, Thorpe, & Siegesmund, 2002) approached the project as an important benchmark for information about arts related opportunities and environments for learning. In this case, assessment is a catalyst for research in and about learning in the arts. However, in the analysis of an earlier NAEP Arts assessment, Brent Wilson voiced a concern that "high school students did reasonably well on expressive and design tasks and that they did poorly—sometimes very poorly—on tasks that require knowledge of the history of art and understanding of the art world." (Council of Chief State School Officers, et.al, 1994, p. 4). For this analysis, the purpose of the assessment was to reveal strengths and weaknesses in high school art programs and curricula.

Another large scale assessment is the College Board's Advanced Placement Program that offers students the opportunity to do college level work while still in high school. Students may take Chemistry, English, Biology, Government, Music Theory, Foreign Language, Calculus, Computer Science, Economics, Art History, and Studio Art and be evaluated nationally through testing and in the case of Studio Art have their portfolios evaluated by a team of expert readers (College Board, 2005). The College Board's AP Studio Art Portfolio consists of a portfolio of significant works of art created by the student following guidelines that have been designed to accommodate a variety of interests and approaches to art. There are three required sections: (1) Quality which reflects the development of a sense of excellence in art; (2) Concentration which demonstrates an in-depth, personal commitment to a particular artistic concern; and (3) Breadth which reveals a variety of experiences in using the formal, technical, and expressive means available to an artist. The resulting scores of students' AP exams and/or portfolios provide means for colleges to grant credit, placement, or both. Many high schools also offer incentives for students and teachers to participate in the advanced placement program such as weighted credits, special seals on their diplomas, additional planning time, and re-certification points through special training workshops.

Figure 10. 4. Jeremy Meisel. *Artist at work* series, digital print, 8 x 10 inches. In this work Meisel explores the process of reflection, evaluation, and assessment through a personal analysis of his everchanging aesthetics.

Teacher Assessment

Although assessment approaches for clarifying what students know and are able to do is directly linked to schools and teachers, there are specific assessment devices for determining how well teachers teach.

National Board Teacher Certification offers teachers opportunities to be evaluated on a "series of performance based assessments that includes teaching portfolios, student work samples, videotapes, and thorough analyses of classroom teaching and student learning" (NBPTS™ 2005, para. 10). The certification is based on four propositions: (1) teachers are committed to students and their learning; (2) teachers know the subjects they teach and how to teach those subjects to students; (3) teachers are responsible for managing and monitoring student learning; and (4) teachers think systematically about their practice and learn from experience. To be eligible for NBTC a teacher must possess a baccalaureate degree from an accredited institution, complete three years of successful school teaching, and hold a valid state teaching license. There are two NBTC certificates for art teachers including the Early and Middle Childhood Certificate (Ages 6-10) and the Early Adolescent through Young Adulthood Certificate (Ages 11-18). Touted as both a professional development experience and rigorous assessment, art teachers applying for NBTC must submit completed forms, critical essays, video tape and analyses of teaching related to five themes: (1) A Portrait of Teaching Over Time; (2) Learning About Making Art; (3) Learning to Study, Interpret, and Evaluate Art; and (4) Documented Accomplishments: Contributions to Student Learning.

The Praxis Series™ are tests that states in the U.S. use as part of their teaching licensing certification process. Praxis I® evaluates basic academic skills; Praxis II® measures general and subject-specific knowledge and teaching skills; and Praxis III® assesses classroom performance (ETS 2005, para 1, retrieved online March 13, 2006 from http://www.ets.org/praxis/). Different states have differing testing and scoring requirements.

Administrative evaluations are formal evaluations typically performed by principals and assistant principals. Completing a rubric outlined on a form and evaluating the teacher's lesson plan, principals typically observe each teacher two to three times a year. Following this observation, the principal writes an official report and meets personally with the teacher to review her or his observations and together work toward a plan for improving and/or sustaining quality instructional strategies. In a similar way, student teacher evaluations involve university supervisors and cooperating teachers in observation and evaluation of teaching performance.

One of the most important forms of evaluating teaching is teacher self-assessment. Teachers are remarkably vigil in this process. They ask their students, invite other teachers to come in their classroom, keep running notes on lesson and unit plans explaining what worked and what didn't. Teaching journals are actively encouraged in many teacher-education programs. One of the most beneficial teaching assessment activities we see is in the form of videotaping. Simply setting up a camera in one area of the room or asking a colleague or student to tape for you and then watching the video or DVD privately or with groups can reveal so much about what is going on in the classroom that teachers often do not see. A teacher may see that she or he did not speak to one particular student during the class or that he or she spent most of her or his time in one area of the room. Student expressions missed during the class may reveal a need for clarification or review. Caution must be taken to destroy the videos after scrutiny as permission to tape for public viewing is necessary and many schools do not allow videotaping in the classroom for security reasons. Although many administrators support this type of self-assessment, we advise that you first ask an administrator in your school before doing any kind of photography or video that include images of students.

Curriculum and School Assessment

As we stated earlier, assessment is a way to make visible what students have learned. A final way to think about assessment is to consider the impact on the overall school program related to interdisciplinary studies. As for the total school curriculum, it can be "assessed in terms of balance, coverage, and appropriateness" (Campbell and Harris, 2001, p. 155). Teachers can contribute to the school-wide assessment with "revisions, additions, and refinements to the curriculum" (Campbell and Harris, 2001, 169) as well as suggestions for new tasks or assignments. They can also "generate school-wide data about all themes taught in the school" (Campbell and Harris, 2001, p. 169). Questions like these could be posed: What happened during the units of instruction? What needs to be changed or modified to improve outcomes, instruction, or facilitation?

A standardized school assessment requirement that has made a huge impact on U.S. schools of late is the *No Child Left Behind Act* (NCLB) of 2001. "In theory the law articulates the idea that all students can learn more than teachers may expect of them. . . In its requirements, the law demands teachers who are well qualified to impart knowledge to students, and seeks greater engagement of parents in monitoring the quality of the schools their children attend" (Chapman, 2005). Although accountability through assessment is of course important for quality schools, the punitive nature of this law has many school officials, teachers, students, and parents crying foul. Basically, school funding is the issue. If schools do not comply with what the act requires as "adequate yearly progress" necessary (95% to 100% of students score proficient or above in reading, mathematics, and science by 2014) they are at risk of losing federal grants and having such negative classifications as "A School In Need of Improvement" or "A School In Need of Corrective Action." The act and its consequences are much too involved to completely review in this chapter. We include it here and again in the Discussion Questions and Activities at the end of this chapter as both an acknowledgment of high-stakes testing and a challenge to teaching art in interdisciplinary ways. What is at issue is that the *No Child Left Behind Act* may have grave consequences that do not place the visual arts at the center of interdisciplinary approaches to high school art as we have promoted in this book. It is critically important now more than ever to develop, administer and report findings from visual art education assessments.

Summary

"What you test is what you get. . . You don't get what you don't test," (Council of Chief State School Officers, et al. 1994, p. 8). So, what does it mean to assess in an interdisciplinary art curriculum? We hope

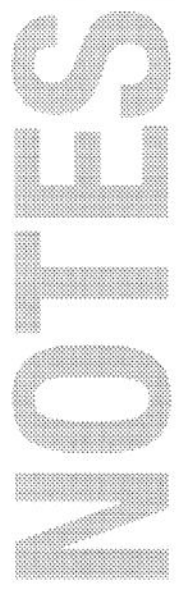

that this chapter has made clear that there are many forms and purposes of assessment. Exhibitions and portfolios can both be used as assessment tools just as easily as essays and performances. Perhaps the bottom line is that assessment serves as a means to provide feedback to teachers, students, parents, administrators, and the public about student learning as well as the degree to which student learning has occurred. The feedback offered through assessment also responds to the degree to which teachers have taught, and curricula were designed and implemented.

> Teachers know that they will be judged by test results, and they will continue to teach to the test by which they are judged. That is why we. . . contend that the tests must test knowledge and skills that are meaningful to students. The tests should be based on what children learn in class, and what they learn in class should be worth knowing and doing and using (Ravitch, 1993, p. 769).

Many curriculum designers spend careful time aligning their expectations with those mandated by national, state, and local standards. But we want to emphasize that these standardized expectations are not curriculum but rather each is a set of expectations that a select group of individuals has assembled as a way to say, "we think these concepts, ideas, and pieces of knowledge are important for students to know, learn, and do at this level." The key task for good inderdisciplinary curriculum designers is to put together meaningful assignments for students that enable them to go beyond and make connections among these standards. We believe that students need to learn more than what is expected of them by such mandated standards.

Assessment is more than a test or quiz that comes at the end of a unit. At its very best, assessment should be directly tied to the objectives and goals of an activity, lesson, unit or curriculum plan. And without a doubt, assessment should be part of the learning experience for students rather than an add-on or after-thought. That is, assessment should drive the design of curriculum rather than be treated as a hitchhiker.

Discussion Questions and Activities

1. Read Laura Chapman's (2005) article "No Child Left Behind in Art?" in *Art Education 58*(1), pp. 6-16. Referring back to the activity you completed at the end of Chapter 1, write a letter to your Congressional representative explaining the importance of visual art in the education of every U.S. citizen.

2. Explain the difference between assessment and evaluation. How will you assess one of your interdisciplinary units of instruction? How will you evaluate a student's work in this unit? From what criteria will you base your assessment and your evaluation?

3. Go to the Praxis web site and download the "Tests at a Glance" that are required in your state. Take the practice tests included in the booklets. How will you study and prepare for these tests? We suggest that you formulate study groups and invite local experts such as high school math and English teachers to assist you in reviewing basic academic skills. Peruse your art textbooks to review art history, foundation, media, technique, and processes.

Notes

[1] For more on Arts Propel and process-folio assessment, see *Taking Full Measure: Rethinking Assessment Through the Arts* (1995) by Dennie Palmer Wolf and Nancy Pistone.

[2] Author Pamela G. Taylor made these observations of Ms. Hinson's practices during many years of working with her both in the Montgomery County Virginia school system and through student teacher placements with Radford University, Radford, VA.

Chapter 11
IMAGINING THE POSSIBILITIES

High school art teachers work diligently to provide their students with meaningful educational experiences in and through the visual arts. After all, art is an essential life skill for all students. [Refer back to Chapter 1.] Yes, some high school students see art class as an opportunity to learn and develop artmaking techniques and strengthen portfolios for application to colleges and art schools. For other students, art class is a hobby or area of interest that they believe to be unrelated to their future dreams. And, there are still other students who take art in high school simply for the curriculum requirements and credits toward graduation. But, no matter what their reasons are for being in art class, their art teachers, like Denise Stanley constantly seek ways to help them successfully face the challenges of their high school experiences.

> During my years as a high school art teacher in Oregon, I have found that students who may have insecurities about their abilities as young artists have discovered successful results when creating artworks through means of design rather than continuously attempting unsuccessfully at creating artworks that are meant to appear realistic in nature. I have also learned that many high school students find mathematics to be one their most challenging academic subjects. Through art, I believe math becomes a more enjoyable subject for these students to learn (Denise Stanley, personal communication, September 22, 2004). [Teacher contribution.]

Approaches and Criteria — Administrative Strategies

Technique and Skill Fundamentals — Themes/Ideas

At a time when educators are witnessing numerous, simultaneous, and sometimes contradictory initiatives to restructure high schools, many teachers may feel pressure to restructure and reassess the content of their curricula with respect to state mandated standards and guidelines. For visual art teachers, the pressure of such demands may be manifested in supportive roles to curricula in "other subject areas" that are viewed as more essential or more important (Keifer-Boyd and Smith-Shank 2006). We are not unfamiliar with these demands and the precarious position in which the visual arts seem to be placed within the curriculum by administrators, legislators, and school board officials. We believe that a strong interdisciplinary approach to curriculum in and through visual art addresses effectively these seemingly omnipresent initiatives for school reform. Many high school teachers may also find themselves making a transition in their planning and instruction. Such transitions are not always the result of recently mandated attention to state education standards, but rather changes in their own approaches to thinking, learning, and teaching art to high school students. Like Ms. Stanley, many teachers find that interdisciplinary approaches to teaching and learning in the art curriculum are useful means toward reaching educational goals. What are the implications related to interdisciplinary approaches to teaching art in high school? What characteristics do meaningful interdisciplinary approaches share? What support can teachers seek or depend on to insure successful interdisciplinary instruction and learning in the high school art curriculum?

In this final chapter we summarize the major issues and implications associated with meaningful interdisciplinary approaches to teaching art in high school. Throughout this book we have used existing and imagined units of instruction as interdisciplinary exemplars. In this final chapter we present specific yet adaptable criteria for interdisciplinary approaches to teaching art in high school. Finally, we suggest ways in which art supervisors, school districts, local and state art education organizations, art museums and art centers, and university art education departments can help support the efforts of high school art teachers interested in accepting the challenge of imagining and redesigning an integrated art curriculum.

Major Issues and Implications

Interdisciplinary approaches to teaching and learning in the art curriculum require expanded ways of thinking, learning, and making that are actually inherent to the visual arts. As art teachers know studying and making visual art is a highly complex endeavor. Artists continually develop and refine their artmaking skills as ways of working through study, practice, awareness and research. They learn from and become attached to historical and contemporary works of art and artists. They experiment, measure, invent, and connect. They relate their ideas to books, news media and other visual culture references that surround them. They correlate what they draw with what they see and hear. They associate and interact with people and places in the world. They travel and correspond. They cooperate and collaborate while at the same time rally around and protest causes in which they believe. They assemble and combine materials to mediate their ideas. They travel and correspond. They talk and listen. And most importantly, they think before, during, and after they make art.

Transferring the complex ways that artists work and think into the high school art classroom may appear obviously interdisciplinary and yet, learning with the arts is different than learning through the arts (Cornett, 2003; Goldberg, 2006). For example, when students learn with the arts, they study an art form or work of art as a means to explore ideas, concepts, or events from a particular time. This approach also allows students to work creatively (Cornett, 2003). [See Chapter 3.]

Learning through the arts means that students create works of art to correlate, depict and convey their understandings and investigations of ideas, concepts, problems, skills, and technique. [See Chapter 2.] "Teaching through the arts involves creating a classroom in which students actually live and learn through the arts" (Cornett, 2003, p. 44). Whether in or through the arts, the interdisciplinary nature of visual art learning experiences is evident through students' engagement with multiple and broad-based ways of thinking as well as doing.

We have also explored the idea of interdisciplinary curriculum as integrated curriculum. [See Chapter 2.] For example, Cornett (2003) observed, "Integration of the arts can occur along a continuum from a small degree, at a surface level, to total arts infusion throughout the curriculum. The latter idea includes a respect for the arts as unique disciplines that connect unique communication modes to study science, social studies, math, and the language arts" (p. 43). Such an approach resembles the "learning with the arts" approach. [See Chapters 7, 8, and 9.]

Artmaking Technique and Skill Fundamentals

In this final chapter, we want to make it very clear that interdisciplinary and integrated teaching and learning in or through visual art is first and foremost art-centered. Although we have discussed and given examples of the ways that interdisciplinary approaches inform and promote learning across the high school curriculum, the primary objective of art teachers and students continues to be teaching and learning about art and artmaking. In spite of everything, visual art students still need to explore and acquire technical and media skills. They need to understand composition, color theory, originality, inventiveness, and artisanship. High school art students should continue to be able to demonstrate their knowledge of such artmaking devices as value, scale, perspective, and proportion in an interdisciplinary approach. The difference is that interdisciplinary art teaching and learning approaches help our students acquire and demonstrate these skills in more meaningful and relevant ways that inform and connect their lives with their world outside of the art class. Like the artists and art that we study in our art classes, students whose art educational

When students learn with the arts, they study an art form or work of art as a means to explore ideas, concepts, or events from a particular time. . . .Learning through the arts means that students create works of art to correlate, depict and convey their understandings and investigations of ideas, concepts, problems, skills, and technique.

Photograph by Donna Green, 2006.

High school art students should continue to be able to demonstrate their knowledge of such artmaking devices as value, scale, perspective, and proportion in an interdisciplinary approach. The difference is that interdisciplinary art teaching and learning approaches help our students acquire and demonstrate these skills in more meaningful and relevant ways that inform and connect their lives with their world outside of the art class.

experiences are integrated with other classes and realms of experiences, use their technical skills to create deep and meaningful works of art.

For this reason, we want to say, yes, an interdisciplinary approach to visual art includes the teaching and learning of drawing techniques, but as exercises to grasp, use, and explore when looking for and creating meaning through art. Just as Judy O'Neal Miller, art teacher at Round Rock High School, Round Rock, Texas involves her students in color and value worksheets [Teacher contribution], an understanding and ability to mix and utilize the expressive qualities of colors and their relationships with each other can be important in an integrated visual arts instruction when applied to the expression and exploration of ideas. What's more, requiring high school art students to continually plan, reflect, write, and yes, draw in their sketchbooks and journals is fundamentally an interdisciplinary activity.

Criteria for Exemplary Interdisciplinary Units

The following criteria are an expanded version of beliefs and assumptions suggested by Jacobs (1989) for teachers to consider as they develop their "philosophy for interdisciplinary work" (pp. 9-10):

• Students should have a range of curriculum experiences that reflects both a discipline-field and an interdisciplinary orientation. For art teachers, relating ideas and ways of working between and among disciplines should be done in tandem with rigorous visual art-specific study. [See Chapters 1-10.]

• To avoid the potpourri problem, teachers should be active curriculum designers and determine the nature and degree of integration and the scope and sequence of the study. Art teachers must take an active and vigilant role in the development and continual assessment of interdisciplinary curriculum. [See Chapter 4.]

• Curriculum making is a creative solution to a problem, hence, interdisciplinary curriculum should only be used when the problem reflects the need to over come fragmentation, relevance, and the growth of knowledge. In other words, we are not promoting the idea of interdisciplinarity for the sake of interdisciplinarity. We are challenging teachers to look for and create learning experiences in which interdisciplinary connections lead to superior ways of knowing. [See Chapters 1-10.]

• Curriculum making should not be viewed as a covert activity. The interdisciplinary unit or course should be presented to all members of the school community. Similarly, all visual art teaching and learning should be celebrated and shared through constant exhibition and promotion. [See Chapter 4, 7, and 10.]

• Students should study epistemological issues. Regardless of the age of students, epistemological questions such as "What is knowledge?", "What do we know?", and "How can we present knowledge in the schools?" can and should be at the heart of our efforts (Jacobs and Borland, 1986).

Similarly, art teachers should always ask themselves "Is this worth knowing?" with each lesson and activity they plan. [See Chapter 2 and 5.]

• Interdisciplinary curriculum experiences provide an opportunity for a more relevant, stimulating, and less fragmented experience for students. For art teachers, the expansive opportunities for interdisciplinary connections include visual culture, digital media, and service-learning. [See Chapters 1, 8, and 9.]

• Students can and, when possible, should be involved in the development of interdisciplinary units. Student initiated inquiry [Chapter 5] takes many forms in an art class that is idea-based [Chapter 2], includes meaningful artmaking experiences [Chapter 3], is culturally sensitive and diverse [Chapter 6], connects with the community [Chapter 7], is technologically relevant [Chapter 8], corresponds to the real visual world around them [Chapter 9], and is evaluated and assessed in valid and explicable ways [Chapter 10].

Strategies to Support Integrated Art Curricula

Throughout the book we have focused on ways in which teachers and students plan and engage in interdisciplinary instruction and learning. While we realize that most learning and teaching in high school actually happens in the interaction between teachers and students, we also acknowledge the importance of other players in the development, facilitation, and support of such curricula. We understand fully the important roles of art supervisors, school districts, local and state art education organizations, and university art education departments and the support they offer to the efforts of high school art teachers who are interested in accepting our challenge of imagining and redesigning an integrated art curriculum. [See Chapter 4.]

Planning Time. Teams of teachers need time to plan effective and meaningful interdisciplinary curriculum. Although we recognize and advocate that such planning time should be included in the school day and year, we urge art teachers to take advantage of such shared teacher to teacher moments as lunch, hall, and bus duty, copying, faculty meetings, sports events, assemblies, and parent-teacher conferences. Listening and encouraging students to share what they are learning and doing in their other classes also provide exciting ideas and opportunities for interdisciplinary teaching collaboration.

Administrative Support. Great ideas from teachers are not often realized when they lack administrative support. Desire to achieve and implement meaningful instruction is limited without such support. The desire and hard work of teachers can be meaningless and in vain without real, honest support of administrators in the school and district. We recommend an open and continual communication between school administration and art teachers. Make sure they know what you are doing and why. Ask for their advice as well as their support. Offer them opportunities rather than problems and ideas rather than complaints. [Chapter 8.]

Instructional Resources. Caution must be taken to not rely completely on manufactured interdisciplinary products. Art teachers are highly creative people whose ideas for interdisciplinary instruction are typically much more meaningful to their students than prepackaged models. That said, we also recognize that there are some very good and exciting resource materials avail-

able that will greatly assist teachers in their quest for interdisciplinary instructional strategies if approached critically as well as creatively.

Professional Development. We have found that the high school principals and school district supervisors who organize professional and staff development sessions and workshops are very open to ideas and suggestions from teachers. Therefore, we suggest strongly that art teachers become proactive in giving ideas, offering the names of workshop leaders, and organizing art centered interdisciplinary professional development opportunities for all teachers. Art teachers may work together to share their own exemplary interdisciplinary work. University and college art and art education faculty are typically honored to lead such workshops. Community and professional service is an important aspect of university faculty positions. In addition, university faculty who are closely associated to schools are better able to prepare the pre-service art teachers in their classes. It is also important to look at the wonderful resources in your own community such as artists, museums, galleries, stores, and personalities.

Idea-based and Thematic Curriculum Approach. Our experiences as teachers, mentors, researchers, collaborators and observers of and with high school art teachers suggests that the most meaningful interdisciplinary curricula in art are those that are designed around the investigation and exploration of ideas and themes. Most often, works of art are interpreted to reveal overarching ideas and themes that form the basis for the essential questions and the focus for units of instruction.

> The value of a theme [and idea] study is enabling learners to move beyond the accumulation of fragmented information and facts to the pursuit of deep understanding of global concepts. Simultaneously, students are learning processes, strategies, and skills that can be transferred to future problem solving. Thus, a theme [or idea] study can be defined as a relevant, in-depth study that makes interdisciplinary links, has a clear central focus, and leads to the construction of transferable knowledge and processes. (Campbell and Harris, 2001, pp. 17-18).

Communication. Teachers must effectively communicate with other members of their team, administration, visiting artists, community members, and other key players in the development and implementation of their interdisciplinary curricula. Communication among the various parties involved in interdisciplinary instruction must occur before, during, and after implementation of curriculum. [See Chapter 4, 8, and 9.]

Flexibility. Teachers must be flexible in the planning of instruction on a variety of levels. When the overall goal is clear, flexibility is easier to achieve than when the goal is less well defined. We also believe strongly that teachers must be comfortable and confident enough to recognize that students themselves may offer an alternative plan or opportunity that is much more relevant and meaningful than what was originally planned. [See Chapter 5.]

Concluding Thoughts

Because high school art has been in existence for a long time, we recognize that high school art teachers have been exposed to many varied approaches, trends, and proposals throughout their careers. From open to closed classrooms and standardized tests to portfolio assessments, one art teacher shared with us that if "you teach long enough, you will see many ideas come back around several times" (Wonderley, personal communication, February 23, 1996). Similarly, an interdisciplinary and integrated approach to teaching and learning is not new to education. It is, however, an approach that has not been formally related to or publicly acknowledged in high school art. Therefore, we wrote this book to recognize as well as inspire, encourage, and support the design and implementation of meaningful interdisciplinary curriculum for high school art. This recognition includes what high school art teachers already do together with some imagined possibilities of what they might consider doing. We hope that this recognition will enable veteran, new, and pre-service high school art teachers to be interdisciplinary leaders in their schools who, like the artists that they are, think, see, teach, learn, and create in artfully connective ways that give meaning to their lives.

REFERENCES

ACA Group (2004). Skills for the creative workforce. Retrieved December 30, 2004 from http://www.theaca group.com/skills_creative_workforce.htm

Adler, M. (1983). *The paideia proposal: An educational manifesto*. New York: MacMillan.

Anderson, T. & Milbrandt, M. (2005). *Art for life*. New York: McGraw-Hill.

Archer, J. (2000). Summit to issue call for service in name of youths. In *Education Week on the Web, 16* (30). Retrieved June 19, 2000 from http://www.edweek.org/ew/vol- 16/30summit.h16

Armstrong, C. L. (1994). *Designing assessment in art*. Reston, VA: National Art Education Association.

Arnheim, R. (1969). *Visual thinking*. Berkeley, CA: University of California Press.

Art Gardens of Pittsburgh (2002). About the gardens. Retrieved August 20, 2005 from http://www.persephone-project.org/ArtGardens.html

Ballengee-Morris, C. (1995). Roots, branches, blossoms: Cultural colonialism. *Working papers*. Reston, VA: NAEA.

Ballengee-Morris, C. (1997). A mountain cultural curriculum: Telling our story. *The Journal of Social Theory in Art Education, 17*(97), 98-116.

Ballengee-Morris, C. (1998). Paulo Freire: A community-based art education. *The Journal of Social Theory in Art Education*, *18*(98), 97-115.

Ballengee-Morris, C. (2000a). Decolonializaltion, art education, and one Guarani nation of Brazil. *Studies in Art Education*. *41*(2), 100-113.

Ballengee-Morris, C. (2000b). A sense of place: Allegheny echoes. In K. Congdon, D. Blandy, P. Bolin (Eds.), *Making invisible histories of art education visible* (pp. 176-187). . Reston, VA: NAEA.

Ballengee-Morris, C. (2006). They came, they saw, they named, and maintain. In L. Emerson (Ed), *Journal of Indigenous Studies*. In print.

Ballengee-Morris, C., & Striedieck, I.M. (1997). A postmodern feminist perspective on visual arts in elemen tary teacher education. In D. R. Walling (Eds.) *The Role of the arts and humanities in postmodern schooling* (pp. 193-215). Bloomington, IN: Phi Delta Kappa Educational Foundation.

Ballengee-Morris, C. & Stuhr, P. L. (2001). Multicultural art and visual cultural education in a changing world. *Art Education*, *54*(4), 6-13.

Ballengee-Morris, C. and Taylor, P. (2005). You can hide but you can't run: Interdisciplinary and culturally sensitive approaches to mask making. *Art Education*, *58*(5), 12-17.

Banks, J. A. & Banks, C. A. (Eds.), (1995). *Handbook of research on multicultural education*. New York: Macmillan Press.

Barrett, T. (1994). Principles for interpreting art. *Art Education, 46*(1), 8-13.

Barrett, T. (2003). Interpreting visual culture. *Art Education, 56*(2), 7-12.

Bateman, W. L. (1990). *Open to Question*. San Francisco: Jossey-Bass.

Bigelow, B., Harvey, B., Karp, S. & Miller, L. (Eds.). (2001). *Rethink ing our classrooms, 2*. Milwaukee, WI: Rethinking Schools, Ltd.

Bloorview Macmillan. (2005). Spiral garden and cosmic bird feeder at the Children's Centre in Toronto, Canada. Retrieved August 26, 2005 from http://www.bloorviewmacmillan.on.ca/Spiral/

Bochner, M. (2004). About the Kraus Campo Garden. Retrieved August 20, 2005 from http://www.cmu.edu/cfa/aboutgarden.html

Bolin, P. E. & Blandy, D. (2003). Beyond visual culture: Seven state ments of support for material culture studies in art education. *Studies in Art Education, 44*(3), 246-263.

Brady, J. (1995). *Schooling young children: A feminist pedagogy for liberatory learning*. Albany, NY: State University of New York Press.

Brewer, M. B. & Miller, N. (1984). Beyond the contact hypothesis: The oretical perspectives in desegregation. In N. Miller and M. B. Brewer (Eds.), *Groups in Contact: The Psychology of Group Desegregation* (pp. 281-302). New York: Academic.

Brown, L. & Craven, J. (1999). Beginning from the end: The Jerusalem architecture project. In D. P. Wolf & D. Balick (Eds.) *Art works! Interdisciplinary learning powered by the arts* (pp. 1-6). Portsmouth, NH: Heinemann.

Brown, N. (2003). Commentary: Are we entering a post-critical age in visual arts education? *Studies in Art Education, 44*(3), 285-289.

Buckingham, M. & Clifton, D. (2001). *Now, discover your strengths*. New York: The Free Press.

Burton, D. (2001). A quartile analysis of the 1997 NAEP Visual Arts report card. *Studies in Art Education, 43*(1), 35-44.

Burton, D. (2006). *Exhibiting student art*. New York: Teachers College Press.

Byron, E. (2005, July 27). To master the art of solving crimes, cops study Vermeer. *The Wall Street Journal* CCXLVI (18), A1-8. Cambridge Arts Council. (2002). Arts on the line: Harvard Square MBTA Station. Retrieved January 6, 2006 from http://www.ci.cambridge.ma.us/~CAC/public_art_tour/map_05_harvard.html.

Campbell, D. M. and Harris, L. S. (2001). *Collaborative theme build ing: How teachers write integrated curriculum*. Boston: Allyn & Bacon.

Canter L. & Canter, M. (1976). *Assertive discipline*. Santa Monica, CA: Lee Canter & Associates.

Carpenter, B. S. (1996). A meta-critical analysis of ceramics criticism for art education: Toward an interpretive methodology. Unpublished doctoral thesis. The Pennsylvania State University.

Carpenter, B. S. (1999). Art lessons: Learning to interpret. *Educational Leadership, 57*(3), 46-48.

Carpenter, B. S. (2003). Never a dull moment: Pat's barbershop as educational environment, hypertext, and place. *The Journal of Cultural Research in Art Education, 21,* 5-18.

Carpenter, B. S. (2005). The return of visual culture. *Art Education, 58*(6), 4-5.

Carpenter, B. S. & Sessions, B. (2002). (Re)Shaping visual inquiry in three dimensional art objects in the elementary school: A content-based approach. In Y. Gaudelius & P. Speirs (Eds.) *Contemporary Issues in Art Education* (pp. 370-382). Upper Saddle River, NJ: Prentice Hall.

Carpenter, B. S. & Taylor, P. G. (2003). Racing thoughts: Altering our ways of knowing and being through computer hypertext. *Studies in Art Education, 45*(1), 40-55.

Cedco Writing Group. (1991). *Contemporary American women artists*. San Rafael, CA: Cedco Writing Group.

Cembalest, R. (1991). The ecological art explosion. *ARTnews, 90*(6), 96-105.

Center for Media Literacy. (2003). About CML. Retrieved September 9, 2005 from http://www.medialit.org/about_cml.html

Chanda, J. & Daniel, V. (2000). ReCognizing works of art: The essences of contextual understanding. *Art Education, 53*(2), 6-11.

Chapman, L. H. (2003). Studies of mass arts. *Studies in Art Education, 44*(3), 230-245.

Chapman, L. (2005). No child left behind in art? *Art Education, 58*(1), 6-16.

Charleston, R. (Ed.). (1969). *World ceramics.* Secaucus, NJ: Chartwell.

Clark, G., & Hughto, M. (Eds.). (1979). *Century of ceramics in the United States.* New York: E.P. Dutton.

Clark, G. & Zimmerman, E. (2004). *Teaching talented art students, principles and practices.* Reston, VA: NAEA.

Clarke, J.H. and Agne, R. M. (1997). *Interdisciplinary high school teaching: Strategies for integrated learning.* Needham Heights, MA: Allyn & Bacon.

College Board. (2005). Studio art. Retrieved October 10, 2005 from http://www.collegeboard.com/student/testing/ap/sub_studioart.html?studioart

Cornett, C. E. (2003). *Creating meaning through literature and the arts: An integration resource for classroom teachers,* (2nd edition). Upper Saddle River, NJ: Merrill Prentice-Hall.

Council of Chief State School Officers, the College Board, & the Coun cil for Basic Education. (Prepared by). (1994). *Arts education assessment framework.* Washington, DC: National Assessment Governing Board.

Daniel, V. A. (2001). Art education as a community act: Teaching and learning through the community. *Conference Proceedings for the International Symposium in Art Education*, Taiwan.

Daniel, V., Stuhr, P. & Ballengee-Morris, C. (no date). Decolonialism, social justice and community-based art education. Unpublished manuscript, The Ohio State University.

Danto, A. (1998). *Beyond the Brillo box: The visual arts in post-histori cal perspective.* Berkeley, CA: University of California Press.

Davidson, R. F. (1952). *Philosophies men live by.* New York: Holt, Rinehart and Winston.

Davis, B., Sumara, D., and Luce-Kapler, R. (2000). *Engaging Minds: Learning and Teaching in a Complex World.* Mahwah, NJ: Lawrence Erlbaum Associates.

Delacruz, E. M. (1997). *Design for inquiry: Instruction, research, and practice in art education.* Reston, VA: The National Art Education Association.

Dewey, J. (1932). The Moral Self (from Ethics). In L. A. Hickman & T. A. Alexander (Eds.), *The Essential Dewey* (1998, vol. 2). Bloomington, IN: Indiana University Press.

Dewey, J. (1938). *Experience and education.* New York: Macmillan.

Diket, R. M. (2001). A factor analytic model of eighth –grade art learn ing: Secondary analysis of NAEP Arts data. *Studies in Art Education, 43*(1), 5-17.

Diket, R. M., Sabol, R. F., Burton, D., Thorpe, P. K., & Siegesmund, R. (2002). Learning from the NAEP Assessment. *Translations in Art Education, 11*(2), 1-6.

Dolev, J. C., Friedlaender, L. K. & Braverman, I. M. (2001). Use of fine art to enhance visual diagnostic skills. *JAMA (Journal of the American Medical Association), 286*(9), 1003-1132.

Doyle, N. (2004). Artist profile: Faith Ringgold. *Nancy Doyle Fine Art.* Retrieved December 16, 2005 from http://www.ndoylefineart.com/ringgold.html

Dubois, P. M. & Hutson, J. J. (1997). *Bridging the racial divide: A report on interracial dialogue in America.* Brattleboro, VT: Center for Living Democracy, Interracial Democracy Program.

Duncum, P. (1999) A case for an art education of everyday aesthetic experiences. *Studies in Art Education, 40*(4), 295-311.

Duncum, P. (2001). Theoretical foundations for an art education of global culture and principles for classroom practice. *International Journal of Education and the Arts, 2*(3). Retrieved January 17, 2004 from http://ijea.asu.edu/v2n3/

Duncum, P. (2002). Clarifying visual culture art education. *Art Education, 55*(3), 6-11.

Duncum, P. (2003). Visual culture in the classroom. *Art Education, 56*(2), 25-32.

Duncum, P. (2004). Visual culture isn't just visual: Multiliteracy, multimodality and meaning. *Studies in Art Education, 45*(3), 252-264.

Durant, W. (1962). *The story of philosophy.* New York, NY: Time.

Drake, S. (1998). *Creating integrated curriculum.* Thousand Oaks, CA: Corwin Press.

Drake, S. & R. Burns. (2004). *Meeting standards through integrated curriculum.* Alexandria, VA: Association for Supervision and Curriculum Development.

Education World (2003). U.S. Education Standards: National Standards. Retrieved September 12, 2005 from http://www.education-world.com/standards/national/index.shtml

Education World (2005). National standards. Retrieved October 2, 2005 from http://www.educationworld.com/standards/

Eisner, E. W. (1998). Does experience in the arts boost academic achievement? *Art Education, 51*(1), 7-15.

Eisner, E. W. (2004). What can education learn from the arts about the practice of education? *International Journal of Education and the Arts, 5*(4). Retrieved January 18, 2005 from http://ijea.asu.edu/v5n4/

Elkins, J. R. (2000). Practical moral philosophy for lawyers: Socrates and the Socratic method. Retrieved July 11, 2005 from http://www.wvu.edu/~lawfac/jelkins/pmpl99/teachvirtue/socrates.html

Empty Bowls. (2000). Empty Bowls: What's new? Retrieved November 23, 2001 from http://www.emptybowls.net/Whatsnew.htm

ETS (Educational Testing Service). (2005) The PRAXIS Series™. Retrieved October 10, 2005 from http://www.ets.org/portal/site/ets/menuitem.435c0b5cc7bd0ae7015d9510c3921509/?vgnextoid=48c05ee3d74f4010VgnVCM10000022f95190RCRD

Eyler, J. & Giles, D. E. (1999). *Where's the learning in service-learning?* San Francisco, CA: Jossey-Bass.

First Nations Art. (1999). James Luna. An introduction to contemporary native artists in Canada. Retrieved January 2, 2006 from http://collections.ic.gc.ca/artists/luna.html

Fogarty. R. (1991). *How to integrate the curricula.* Arlington Heights, IL: SkyLight Professional Development.

Fowler, C. (1996). *Strong arts, strong schools.* New York: Oxford University.

Fox, M. L. (2003). Creative thinking for business: Sly as a Fox. Online. Retrieved December 30, 2004 from http://www.slyasafox.com/book/book_1.html

Fransecky, R. & Debes, J. (1972). *Visual Literacy: A way to learn, a way to teach.* Washington, D.C.: Association for Educational Communications and Technology.

Freedman, K. (1999). Reconsidering critical response: Student judgments of purpose, interpretation, and relationships in visual culture. *Studies in Art Education, 40*(2),128-142.

Freedman, K. (2000) Social perspectives on art education in the US: Teaching visual culture in a democracy. *Studies in Art Education, 41*(4), 314-329.

Freedman, K. (2003). *Teaching visual culture: Curriculum, aesthetics, and the social life of art.* New York: Teachers College Press and Reston, VA: NAEA.

Freire, P. (1973). *Education for critical consciousness.* New York: The Seabury Press.

Freire, P. (1994). *Pedagogy of the oppressed.* New York: Continuum. (Original work published in 1970).

Freire, P. (1998a). *Pedagogy of freedom.* Oxford, England: Rowman & Littlefield Publishers, Inc.

Freire, P. (1998b). *Teachers as cultural workers: Letters to those who dare teach.* Boulder, CO: Westview Press.

Furco, A. (2001). Remarks at the First Annual International Conference on Service-Learning Research, Berkeley, CA. October 21-23, 2001.

Gablik, S. (1991). *The reenchantment of art.* New York: Thames and Hudson, Inc.

Geertz, C. (1973). I*nterpretation of cultures.* New York: Basic Books.

Giroux, H. (1992). *Border crossings: Cultural workers and the politics of education.* New York: Routledge.

Goldberg, M. (2006). *Integrating the arts: An approach to teaching and learning in multicultural and multilingual settings,* (3rd edition). Boston: Pearson.

Gollnick, D. M. & Chinn, P. C. (1993). *Multicultural education in a pluralistic society* (4th edition). Columbus, OH: Charles E. Merrill Publishing Company.

Goodwin, M. A. (1993). *Design standards for school art facilities*. Reston, VA: NAEA

Gengnagel, J. (2001). A virtual mirror of KÁDÍ (Kashi). Retrieved Sep tember 22, 2005 from http://benares.uni-hd.de/introduction.htm

Gude, O. (2004). Postmodern principles: In search of a 21st century art education. *Art Education, 57*(1) 6-14.

Gunn, B. & King J. (2003, March). Growth of an interdisciplinary teaching team. *Urban Education Journal, 38*(2), 173-195. Retrieved December 27, 2005 from http://www.wcer.wisc.edu/news/coverStories/interdisciplinary_teaching_team.php

Hall, S. (1992). Cultural identity in question. In S. Hall, D. Held, & T. McGrew (Eds.), *Modernity and its futures* (pp. 273-316). Cam bridge: Polity Press.

Hall, S. (2004). Who needs identity? In P. du Gay, J. Evans, & P. Red man (Eds.), *Identity: Reader* (pp. 15-30). London: Sage Publications, Ltd.

Harrison, H.A. (2004). ART: Unforeseen Irises and a Meteor Asking Big Questions. *New York Times Online*. Retrieved January 6, 2006 from http://query.nytimes.com/gst/fullpage.html?res=9D05EEDA123EF933A05756C0A9629C8B63.

Heffernan, K. (2001). *Fundamentals of service learning course construction*. RI: Campus Compact.

Heise, D. (2004). Is visual culture becoming our canon of art? *Art Education, 57*(5), 41-46.

Henry, C. (1990). Grading student artwork: A plan for effective assessment. In B. Little (ed.) *Secondary Art Education: An Anthology of Issues,* (pp. 61-68). Reston, VA: NAEA.

Henry, C. (1992). Evaluation of student progress in the elementary art classroom. In A. Johnson (Ed.), *Art Education: Elementary*, (pp.201-212). Reston, VA: NAEA.

High school classroom management (2004). Retrieved July 30, 2005 from http://users.pandora.be/education/24.htm

Horton, M. & Freire, P. (1991). *We make the road by walking: Conversations on education and social change*. Philadelphia: Temple University.

Howard, J. (1993). Community service-learning in the curriculum. In J. Galura and J. Howard (Eds.) *Praxis I*, (pp. 3-12). Ann Arbor, Michigan: OCSL Press.

International Visual Literacy Association (2005). Visual Literacy. Re trieved July 18, 2005 from http://www.ivla.org/organization/whatis.htm

Jacobs, H. (1989). *Interdisciplinary curriculum: Design & implementation*. Alexandria, VA: Association for Supervision and Curriculum Development.

Jacobs, H. (1997). *Mapping the big picture*. Alexandria, VA: Associa tion for Supervision & Curriculum Development.

Jacobs, H. H. (1998). *Mapping the big picture: Integrating curriculum & assessment, K-12*. Alexandria, VA: Association for Supervision & Curriculum Development.

Jacobs, H.H., & Borland, J.H. (Winter, 1986). The interdisciplinary concept model: Design and implementation. *Gifted Child Quarterly*.

Jeffers, C. (2005). *Spheres of possibility: Linking service-learning and the visual arts*. Reston, VA: The National Art Education Associa tion.

Jenkins, H. (1992). *Textual Poachers: Television Fans & Participatory Culture (Studies in Culture and Communication)*. New York: Routledge.

Johnson, H. (1998). Class Size. (ERIC Document Reproduction Service No. ED 429349).

Johnson, P. (2004). Arts and technology as the hub for all disciplines. *Tech-Learning*. Retrieved July 17, 2005 from http://www.techlearning.com/story/showArticle.jhtml?articleID189-2826.

Johnston, P. (1985). *Joyce Kozloff: Visionary ornament*. Boston: Bos ton University Arts and Publications.

Johnson, R. T., & Johnson, D. W. (1986). Action research: Cooperative learning in the science classroom. *Science and Children, 24,* 31-32.

Kafai, Y. and Resnick, M. (Eds). (1996). *Constructionism in Practice: Designing, Thinking, and Learning in a Digital World.* Mahwah, NJ: Lawrence Erlbaum Associates.

Kangas, M. (1980). Patti Warashina: The ceramic self. *American Craft, 40*(2), 2-7, & 83.

Kashi. (2005a). About Kashi. Retrieved September 19, 2005 from http://www.Kashi.com/aboutKashi.aspx?SID=1&Category_ID=32&

Kashi. (2005b). Mighty Bites: A Mighty Beginning. Retrieved September 19, 2005 from http://www.Kashi.com/mightybites.aspx?SID=1&Category_ID=77

Kashi. (2005c). Mighty Bites Honey Crunch Cereal. Retrieved September 19, 2005 from http://www.Kashi.com/mightybites_honey crunch.aspx?SID=?SID=1&Category ID=77&LinkID=96&

Kashi. (2005d). The Kashi name. Retrieved September 23, 2005 from http://www.kashi.com/ourname.aspx?SID=1&Category_ID=32&Link_ID=59&

Keifer-Boyd, K. (2003). Three approaches to teaching visual culture in K-12 school contexts. *Art Education, 56*(2), 44-51.

Keifer-Boyd, K. & Smith-Shank, D. (2006). Speculative fiction's con tribution to contemporary understanding: The handmaid art tale. *Studies in Art Education, 47*(2), 139-154.

Kemp, M. (2000). *Visualizations: The Nature Book of Art and Science.* Berkeley, CA: University of California.

Kindler, A. M. (2003). Commentary: Visual culture, visual brain, and (art) education. *Studies in Art Education, 44*(3), 290-296.

Kingery, W. (Ed.). (1984-2000). *Ceramics and civilization* (Vols. 1-9). Westerville, OH: American Ceramics Society.

Krug, D. H. (2003). Symbolic culture and art education. *Art Education, 56*(2), 13-19.

Lampert, N. (2006). Critical thinking dispositions as an outcome of art education. *Studies in Art Education, 47*(1), 215-228.

Lesman A. & Morrow, M. (1999). Report on service-learning faculty fellow projects. VA COOL Symposium on integrating service with the curriculum, Richmond, VA. Retrieved June 28, 2000 from http://www.richmond.edu/~vacool/9899ffblurbs.html

Lewis, D. (1991). *Warren MacKenzie: An American potter*. Tokyo: Kodansha.

Levin, E. (1988). *The history of American ceramics*. New York: Abrams.

Levin, M. (2001). *Teach me!: Kids will learn when oppression is the lesson*. New York: Rowman & Littlefield.

Lynn, M. (1990). *Clay today*. Los Angeles: Chronicle Books.

Májozo, E. C. (1995). To search for the good and make it matter. In S. Lacy (Ed.), *Mapping the terrain: New genre public art*, (pp. 88-93). Seattle: Bay Press.

Maryland Student Service Alliance. (2002). Service-learning project ideas: Caring through communication technology. Retrieved September 6, 2002 from http://www.mssa.sailorsite.net/mp comm.html

Matilsky, B. (1992). *Fragile ecologies*. Queens, NY: Rizzoli.

McBride, K. (2004, January 26). The fine art of the space age: Little-known NASA program documents history through various media. Online Retrieved December 30, 2004 from http://www.washingtonpost.com/ac2/wp-dyn?pagename=article&contentId=A47125-2004Jan25¬Found=true

McPhee, K. (1999, May). The art of numbers. *Plus Magazine*, Issue 8. Online. Retrieved December 30, 2004 from http://plus.maths.org/issue8/features/art

McTighe, J. & Wiggins, G. (2004). *Understanding by design*. Alexandria, VA: Association for Supervision and Curriculum Develop ment.

Miller, N. & Harrington, H. J. (1992). Social categorization and inter group acceptance: Principles for the design and development of cooperative learning teams. In R. Hertz-Lazarowitz & N. Miller (Eds.), *Interaction in cooperative groups: The theoretical anatomy of group learning* (pp. 203-227). New York: Cambridge University.

Mirzoeff, N. (1999). *Introduction to Visual Culture*. London and New York: Routledge.

NASA (National Aeronautics and Space Act) (1958). Retrieved Decem ber 30, 2004 from http://www.hq.nasa.gov/office/pao/History/amendact.html

NBPTS™ (National Board for Professional Teaching Standards. (2005). About NBPTS™. Retrieved October 10, 2005 from http://www.nbpts.org/about/index.cfm

NSES. (1995). National science education standards. National Academy of Sciences. Retrieved January 10, 2002 from http://search.nap.edu/readingroom/books/nses/html/6e.html

Nyman, A. L. & Jenkins, A. M. (1999). *Issues and approaches to art for students with special needs*. Reston, VA: NAEA.

Osbeck, L. M., Moghaddam, F.M. & Perrault, S. (1997). Similarity and attraction among majority and minority groups in a multicultural context. *International Journal of Intercultural Relations, 21*(1), 113-1123.

Osterland, R. (2005). Music exercises. Early music Chicago. Retrieved January 6, 2006 from http:://earlymusichicago.org/sheet_music_exercises.htm

Parks, N. S. (2004). Bamboozled: A visual culture text for looking at cultural practices of racism. *Art Education, 57*(2), 14-18.

Pauly, N. (2003). Interpreting visual culture as cultural narratives in teaching education. *Studies in Art Education, 44*(3), 264-284.

PBS (Public Broadcasting System). (2005). Power and contemporary art. Art: 21. Retrieved January 7, 2006 from http://www.pbs.org/art21/series/seasonthree/power.html#cai

Perelman, L. J. (1993). School's out - public education obstructs the future: Would you send your kid to a soviet collective? Wired 1.1. Retrieved July 15, 2005 from http://www.spinnaker.com/VILA/bbs-docs/docs/wire-ed.txt

Peterson, P. L. & Swing, S. R. (1982). Beyond time on task: Students' reports of their thought processes during classroom instruction in W. Doyle & T. L. Good (Eds.), *Focus on Teaching: Readings from the elementary school journal*, (pp. 228-238). Chicago, IL: The University of Chicago Press.

Posner, K. (2002). *Leadership the challenge. San Francisco, CA: Jossey-Bass.*

Postrel, V. (2003, July). The aesthetic imperative: Why the creative shall inherit the economy. *WIRED Magazine,* 11(07). Retrieved December 30, 2004 from http://www.wired.com/wired/archive/11.07/view.html?pg=1

Ravitch, D. (1993). Critical issues in the Office of Educational Research and Improvement. In J. J. Jennings (Ed.) *National Issues in Education: The Past is Prologue* (pp.151-163). Bloomington, IN: Phi Delta Kappa International.

Resnick, L., Briars, D., and Lesgold, S. (1992). Certifying accomplishments in mathematics: The new standards examining system. In I. Wirsup and R. Streit (Eds.), *Proceedings of the University of Chicago School mathematics project international conference on mathematics education, vol. 3—Developments in school mathematics around the globe,* (pp. 186-207). Reston, VA: National Council of Teachers of Mathematics.

Risatti, H. (1994).Merle Laderman Ukeles, Methanogenesis Project. *The Mountain Lake Workshop: Artists in Locale*, pp. 30-36. The Anderson Gallery, Virginia Commonwealth University, in association with The Mountain Lake Workshop of the Virginia Tech Foundation.

Rockriver, S. (2004). Thermal formations. *Ceramics Monthly, 53*(3), 42-47.

Romero, M. (2003). Face-to-Face: Chinese ancestors and Warhol superstars share a point of view. *Carnegie Magazine*. Retrieved September 23, 2005 from http://www.carnegiemuseums.org/cmag/bk_issue/2003/janfeb/feat1.html

Roukes, N. (1988). *Design synectics: Stimulating creativity in design.* Worcester, MA: Davis Publications.

Sabol, R. (2001). Regional findings from a secondary analysis of the 1997 NAEP Art assessment based on responses to creating and responding exercises. *Studies in Art Education, 43*(1), 18-34.

Saltmarsh J. (Fall 1996). Education for critical citizenship: John Dewey's contribution to the pedagogy of community service learning. In *Michigan Journal of Community Service-Learning*. (Vol. 3, pp. 22-30). Ann Arbor, Michigan: OCSL (Office of Community Service Learning, The University of Michigan).

Sanders-Bustle, L. (2005). Overcoming the challenges of an integrated approach to art education. *NAEA Advisory*, (summer).

Service Learning Center at VA Tech. (2000). Putting knowledge to work for the community. Retrieved September 6, 2002 from http://www.majbill.vt.edu/SL/index.html

Sessions, B. (1996). Ceramics curriculum: What has it been? What could it be? *Art Education, 52*(5), 6-11.

Sessions, B. (1998). A new case for clay: Multi-dimensional high school ceramics education. *Dissertation Abstracts International, 59*(6A), 9836761.

Sherman, N. (2001). Fine art frames diagnosis: Health scout, value added benefits. Retrieved December 30, 2004 from http://jama.ama-assn.org/cgi/content/extract/286/9/1020

Siegesmund, R., Diket, R. M., & McCulloch, S. (2001). Re-visioning NAEP: Amending a performance assessment for middle school art students. *Studies in Art Education, 43*(1), 45-56.

Simmons, S., & Winner, E. (1992) (Eds.). *Arts propel: A handbook for visual arts*. Princeton NJ: Educational Testing Service and the president and fellows of Harvard College (on behalf of Project Zero, Harvard Graduate School of Education).

Sizer, T. (1984). *Horace's compromise: The dilemma of the American high school*. Boston: Houghton Mifflin.

Sizer, T. R. (1993). *Horace's school: Redesigning the American high school*. Boston: Houghton Mifflin.

Sleeter, C. & Grant. C. (1988). An analysis of multicultural research in the United States. *Harvard Educational Review, 57*(4), 421-445.

Smith-Shank, D. L. (2003). Lewis Hines and his photo stories: Visual culture and social reform. *Art Education, 56*(2), 33-37.

South, H. (2004). Introduction to perspective drawing. About.com. Retrieved January 2, 2006 from http://drawsketch.about.com/library/weekly/aa021603a.htm

Stockrocki, M. (Ed.) (2005). *Interdisciplinary art education: Building bridges to connect disciplines and cultures*. Reston, VA: NAEA.

Stoner, B. (executive producer). (1999). Picasso's artistic process. Guernica. *Treasures of the Western World*. PBS (Public Broadcast System) and Stoner Productions, Inc. Retrieved September 20, 2005 from http://www.pbs.org/treasuresoftheworld/a_nav/guernica_nav/gnav_level_1/2process_guerfrm.html

Stuhr, P. (1995). A social reconstructionist multicultural art curriculum design: Using the powwow as an example. In R.W. Neperud (Ed.), *Postmodernism*,(pp. 193-221). New York: Teachers College Press.

Sullivan, G. (2003). Editorial: Seeing visual culture. *Studies in Art Education, 44*(3), 195-196.

Susi, F. (1995). *Student Behavior in Art Classroom: The Dynamics of Discipline*. Reston, VA: National Art Education Association.

Tatum, B. (1992). Talking about racism: An application of racial identity and development theory in the classroom. *Harvard Educational Review, 62*(1), 1-24.

Tavin, K. & Anderson, D. (2003). Teaching (popular) visual culture: Deconstructing Disney in the elementary art classroom. *Art Education, 56*(3), 21-23/32-35.

Tavin, K. M. (2003). Wrestling with angels, searching for ghosts: Toward a critical pedagogy of visual culture. *Studies in Art Education, 44*(3), 197-213.

Tavin, K. (2004). Art education and visual culture in the age of globalization. *Art Education, 57*(5), 47-52.

Taylor, P. G. (1997). It all started with the trash: Taking steps toward sustainable art education. *Art Education, 50*(2), 13-18.

Taylor, P.G. (1999). Hypertext-based art education: Implications for liberatory learning in high school. Unpublished doctoral dissertation. The Pennsylvania State University.

Taylor, P. G. (2000). Madonna and hypertext: Liberatory learning in art education. *Studies in art education, 41*(4), 376-389.

Taylor, P. G. (2002a). *Amazing grace: The lithographs of Joseph Norman*. Boston: The Center for Afro-American Artists.

Taylor, P.G. (2002b). Singing for someone else's supper: Service-learning and Empty Bowls. *Art Education, 55*(4), 6-12.

Taylor, P. G. (2002c). Differences are not worlds apart: Challenging the notion of "fitting in" with the art of Adrian Piper. In B. Young & M. Erickson (Eds.) *Multicultural artworlds: Enduring, evolving, and overlapping traditions,* (pp. 87-90). Reston, VA:

National Art Education Association.

Taylor, P. G. (2002d). Service learning as postmodern art and pedagogy. *Studies in art education. 43*(2), 124-140.

Taylor, P.G. (2003). Target Practice: Take this Take that!: Exploring the power of words through the art of Joseph Norman. Instructional Resource in *Art Education, 56*(3), 25-32.

Taylor, P. G. (2004a). Service-learning Basics. *NAEA Advisory*, Spring.

Taylor, P. G. (2004b). Hyperaesthetics: Making sense of our technomediated world. *Studies in Art Education, 45*(4), 328-342.

Taylor, P. G. (2005). Music video in high school. In G. Szkeley and I. Szkeley (Eds.), *Video art for the classroom*, (pp. 109-118). Reston, VA: National Art Education Association.

Taylor, P.G. & Ballengee-Morris, C. (2003). Using visual culture to put a contemporary "fizz" on the study of Pop Art. *Art Education. 56*(2), 20-24.

Taylor, P. G. & Ballengee-Morris, C. (2004). Service-learning; A language of "we." *Art Education, 57*(5), 6-12.

Taylor, P.G. & Carpenter, B.S. (2002). Inventively linking: Teaching and learning with computer hypertext. *Art Education, 55*(4), 6-12.

Taylor, P. G. & MacDonald, B. (1996). Is Madonna the Warhol of the 90's: Bringing popular culture into the art education classroom. Presentation at the National Art Education Association Convention in San Francisco.

Thompson, K. M. (1995). Maintaining artistic integrity in an interdisciplinary setting. *Art Education, 48*(6), 39-45.

Thompson, J. G. (1998). *Discipline survival kit for the secondary teacher.* West Nyack, NY: The Center for Applied Research in Education.

Ulaby, N. (2004). From microscopes to large-scale sculpture. National Public Radio. Retrieved February 10, 2006 from http://www.npr.org/templates/story/story.php?storyId=4079067

U.S. Department of Education (1997). Seven Priorities of Education: Priority Four. Retrieved October 2, 2005 from http://www.ed.gov/updates/7priorities/part6.html

U.S. Department of Education (2005). Federal role in education. Retrieved October 2, 2005 from http://www.ed.gov/about/overview/fed/role.html?src=ln

U.S. Government. (1999). Copyright. Retrieved July 19, 2005 from http://www.copyright.gov/fls/fl102.html

VA COOL. (2000). Virginia Campus-Community Corps: An Ameri Corps program. Retrieved November 18, 2002 from http://www.richmond.edu/~vacool/vcccintro.html

Villeneuve, P. (2003). Why not visual culture? *Art Education, 56*(2), 4-5.

Walker, S. (2001). *Teaching meaning in artmaking*. Worcester, MA: Davis Publication.

Wass, D. (2004). Cutting edge: Going beyond reality. EdTech, Winter, 41. Retrieved July 17, 2005 from http://edtech.texterity.com/archives/200411/

Wass, D. & Cady, K. (2005) Perspective drawing using 3D software. Adobe Education. Retrieved July 17, 2005 from http://www.adobe.com/education/curriculum/exchange/persp_drawing.html

Weiler, K. (1988). *Women teaching for change: Gender class & power.* New York: Bergin & Garvey Publishers.

Weschler, L. (2002) Robert Irwin Getty garden. Los Angeles: J. Paul Getty Museum.

Wendl, N. (2004). Anderson Gallery to feature Kendall Buster's Model City. Drake University. Retrieved February 10, 2006 from http://www.drake.edu/newsevents/releases/nov04/111004andersongallery.html

Wiggins, G. & McTighe, J. (1998). *Understanding by Design*. Alexandria, VA Association for Supervision & Curriculum Development.

Wilson, B. (1997). *The quiet evolution*. Los Angeles: Getty.

Wilson, B. (2003). Of diagrams and rhizomes: Visual culture, contemporary art, and the impossibility of mapping the content of art education. *Studies in Art Education, 44*(3), 214-229.

Wilson, B., with Feldman, E., Chapman, L., & MacGregor, R. N. (1996). Criticism in art education. Distinguished fellows forum at the National Art Education Association Conference in San

Francisco, CA, March 23, 1996.

Wilson, M. (1998). The text, the intertext and the hypertextual: A story. Keynote address delivered to the Virginia Art Education Association Convention.

Wisconsin Department of Public Instruction. (2002). Bright beginnings and family community school partnership. Retrieved September 6, 2002 from http://www.dpi.state.wi.us/dpi/dlcl/bbfcsp/slc tgpge.html

Wolf, D. & Balick, D. (1999). *Art works! Interdisciplinary learning powered by the arts.* Portsmouth, NH: Heinemann.

Wolf, D. P. & Pistone, N. (1995). *Taking Full Measure: Rethinking Assessment Through the Arts.* New York: College Board.

Woodfill, J. (2004). National space art homepage. Retrieved December 30, 2004 from http://vesuvius.jsc.nasa.gov/er/seh/spaceart.html

Worthen, B. R. (1993, February). Critical issues that will determine the future of alternative assessment. *Phi Delta Kappan*, 444-454.

ABOUT THE AUTHORS

Authors Christine Ballengee-Morris, B. Stephen Carpenter II, Pamela G. Taylor and Billie Sessions at the entrance to the Pennsylvania State University during the Visual Culture of Childhood Symposium in November 2004.

To let you, the reader know a little more about us, the authors, we include our professional biographies on the following pages. As we said in this book's Acknowledgements, we are, by all accounts, four art teachers who share a similar passion and dedication to the field of art education, as do other teachers of visual art. We met in the 1990s during our respective graduate studies at the Pennsylvania State University. We each were keenly aware and affected by the intense passion, dedication and demand for excellence in art education demonstrated daily by our Penn State community of professors and students. Humbled and inspired by our graduate experiences, we each embarked on personal journeys to support and influence art education experiences in varied regions of the U.S. and the world. Throughout our travels, teaching, and research experiences we have come to rely on each other's expertise to assist, guide, and often critically reflect and evaluate. We approached the writing of this book in the same way that we approach our personal and professional relationships— with an enduring trust that though patient and respectful is always fervently honest.

Author Pamela G. Taylor is Chair and Associate Professor of the Department of Art Education in the School of the Arts at Virginia Commonwealth University in Richmond, Virginia. She earned her Ph.D. from the Pennsylvania State University in 1999 and was honored with a J. Paul Getty Doctoral Dissertation Fellowship for her study "Hypertext-based Art Education: Implications for Liberatory Learning in High School." Her research has been published in such journals as *Art Education, Studies in Art Education, Journal of Cultural Research, Innovate, International Education Journal, Computers in Schools, Journal of Educational Multimedia and Hypermedia,* and *FATE in Review*. Dr. Taylor has contributed to several anthologies published by the National Art Education, Canadian Society for Education through Art, and the American Education Research Associations. She authored *Amazing Grace: The Lithographs of Joseph Norman* and her illustrations have been included in books, newspapers, and magazines. She has presented papers at regional, national, international, and online conferences and conventions. (see http://breeze.uliveandlearn.com/p68077584/). She is currently Editor of *Art Education*, the Journal of the National Art Education Association (NAEA) after serving on the editorial review board for four years. Dr. Taylor taught art in public high and middle schools for nine years, public and private elementary schools for three years, pre-school for three years, senior citizen centers for two years, and began teaching at the university level in 1997. A leader in art-based service-learning research and practice, Dr. Taylor has been steeply involved in Empty Bowls and worked with numerous service-learning partnerships in the U.S. and Cuba. Crediting her research /practice with interactive computer hypertext for her collaborative and linking sensibility, Dr. Taylor is known for drawing people together on a number of communal group projects such as the writing of this book. Her research interests combine and correlate interactive digital technology, visual culture, curriculum, and criticism through such topics as music video, assessment, hyperaesthetics, and pentimenti. Making art is both a luxury and a passion for Dr. Taylor whose work ranges from portraiture to mixed-media layered traditional and digital montages dealing with feminist and social learning issues.

Author B. Stephen Carpenter, II, teaches curriculum development, visual culture pedagogy, and creative inquiry through the arts courses in the Department of Teaching, Learning and Culture at Texas A&M University. Dr. Carpenter was an art teacher in public special education, elementary, and high schools and later taught art criticism, ceramics, and drawing to students in art centers, community programs, and universities. His research explores art education and visual culture, hypertext curriculum theory and design, cultural studies through visual inquiry, and ceramics criticism and education. He has presented his work at numerous regional, national, and international conferences. His research has been published in *Art & Antiques, Art Education, Ceramics: Art and Perception, Educational Leadership, The Journal of Cultural Research in Art Education, The Journal of Educational Multimedia and Hypermedia, Studies in Art Education* and *Studio Potter*. His ceramics, mixed-media assemblages, installations, and performance artworks address issues of social justice and critique historical, cultural, and political constructs. In 2001 and 2003 his work appeared in the 3rd and 4th Biennale Internazionale dell'arte contemporanea (Florence Biennale) in Florence, Italy. Dr. Carpenter is past-editor of *Art Education*, the Journal of the National Art Education Association (NAEA), an assistant editor for the *Journal of Curriculum and Pedagogy*, and a member of the editorial board of the *Journal of Cultural Research in Art Education.* He is also co-editor of *Curriculum for a Progressive, Provocative, Poetic and Public Pedagogy*, selected papers from the 6th annual Curriculum and Pedagogy Conference. Dr. Carpenter has received several awards for his scholarship and professional service. In 2000, he was named Southeast Region Higher Education Art Educator of the Year by NAEA, and in 2004 he received the Dorothy Liskey Wampler Distinguished Professorship from the Department of Art and Art History at James Madison University.

Author Christine Ballengee-Morris is Ethnic Arts Coordinator, Graduate Chair, and Associate Professor of the Department of Art Education at

The Ohio State University. She teaches art education classes that specialize in diversity explorations such as Visual Culture and Indigenous Images, Pow Wow 101, and Crash(ING) Race and Gender. From 2001-05 she was the founding director of The Multicultural Center at OSU. This new unit changed how diversity was defined, represented, and supported for students, faculty and community. She has published in journals including the Journal *of Art Education*, *Journal of Social Theory in Art Education*, *Journal of Cultural Studies*, and *Studies in Art Education*. She has written for several anthologies published by the National Art Education Association and Phi Delta Kappa. She serves on several editorial boards such as Studies in Art Education and Journal of Cultural Studies. She is past president of the United States Society for Teaching through Art. She has presented papers at regional, national, and international conferences. Dr. Ballengee-Morris's teaching experiences include fourteen years in the public school system, artist-in-residencies in the public schools in five countries, undergraduate and graduate level courses, and international teaching. She is trained to lead Social Justice workshops and mediation and has served in those positions for several companies, universities, departments, and community organizations. She is a Cherokee-Appalachian and performs flatfoot dance with her musician husband, David Morris, and son, Jack Morris (aka Jeremy Meisel). She has been recognized by her peers with the National Art Education Grigsby Award (research in and commitment to diversity); OSU-Newark research and service award; NAACP Licking County, Ohio's Young Native American Woman leadership award; and Pennsylvania University's teaching award. Ballengee Morris' research interest include self-determination, identity development, integrated curricula, service-learning , visual culture, and arts-based research.

Author Billie Sessions is Professor of Art Education at California State University, San Bernardino and taught junior and senior high school art and ceramics for over 20 years in Wyoming. She is one of a few researchers in the field involved in the interdisciplinary aspects of ceramics education as witnessed by her 1998 doctoral research at Penn State on *A New Case for Clay: Multi-Dimensional High School Ceramics Education*. She has been an NAEA Vice President for the Pacific Region, and has received awards for Higher Education Art Teacher of the Year for California, NAEA Secondary Art Teacher of the Year for the Pacific Region, and served as President of the Wyoming Secondary Art Education Association. She currently serves on the Editorial Review Board for *Art Education*, the Board of Trustees for The National K-12 Ceramic Exhibition at NCECA (National Council for Education in the Ceramic Arts), and on the Advisory Board for the American Museum of Ceramic Art. She has been the recipient of a Fulbright Scholarship, National Endowment for the Arts and National Endowment for Humanities grants, and a J. Paul Getty Doctoral Dissertation Fellowship for her research on contemporary ceramics curriculum. She exhibits her ceramics frequently and has been invited to present papers on Marguerite Wildenhain (a significant historic ceramics teacher and potter) and ceramics education at numerous state, regional and national art and ceramics education conferences. Her primary art education research interests are the interdisciplinary aspects of ceramics education and translating and integrating theory, practice and standards for contemporary ceramics education. She has been published in several anthologies, *Art Education* and *The NCECA Journal*. She has written about historic and contemporary ceramic artists for *Ceramics Monthly* and *Ceramics Art and Perception*. She actively researches and publishes in texts and journals on Marguerite Wildenhain, and curated the exhibition with noteworthy catalog *RIPPLES: Marguerite Wildenhain and Her Pond Farm Students* that traveled from California to New York in 2002. In 2005 she co-curated the exhibition, Women's 'Werk': The Dignity of Craft " featuring Marguerite Wildenhain and Susi Singer two mid-century European ceramic artists. Each spring, she organizes and sponsors a high school Clay Day at CSUSB.

INDEX

A

B

P

Q

R

S

T

AFTERWORD

by Howard Risatti

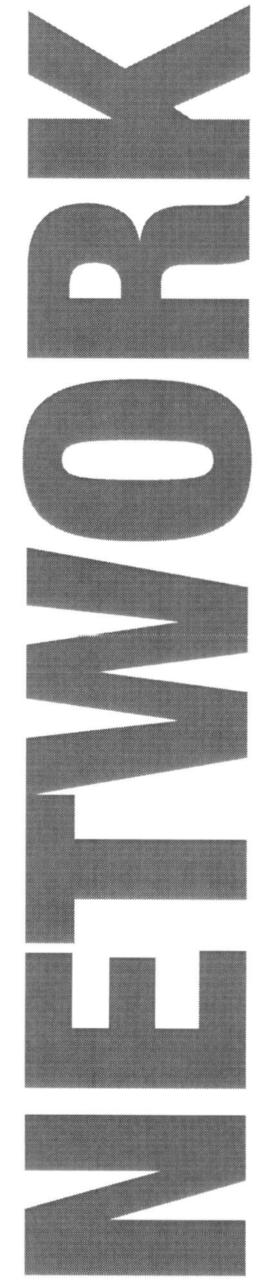

We are honored by the following reflections and contemplations by art historian Dr. Howard Risatti concerning this book. Risatti's work in art history, aesthetics, and postmodern theory have been an inspiration for us throughout our careers. We trust that readers will be both encouraged and inspired by the support and respect for K-12 art teaching and learning that Risatti shares in this Afterword.

Many years ago while a graduate student in Chicago I supported myself, in part, by teaching art and music appreciation in a Catholic high school for girls. I say "in part" because the salary was so low it seemed like part-time pay even though it was full-time work, and hard work at that. Apparently things have changed little over the years, at least in terms of work load and salary. But for me, as I suspect of most teachers, it never was about the pay otherwise we all would have followed the money and become lawyers or stock brokers or business executives or whatever. In my experience people who go into the teaching profession do so because of a passion for ideas, a love of learning, and the satisfaction of watching students mature as human beings. This is certainly true of the two authors of this book who have been my colleagues, Pam Taylor and Steve Carpenter. But I believe it also true of Christine Ballengee-Morris and Billie Sessions for together they write in unison with a clarity and even a sense of urgency that attests to their care and their concern for students and the current state of education in the arts.

Care, concern, and good intentions, in and of themselves, are admirable but not enough to institute change. They must be coupled with a sound strategy and plan of action that is reasonable, economically feasible, and promising in the results it offers. I think what the authors outline in *Interdisciplinary Approaches to Teaching Art in High School* is just such a plan. And while what they propose is not totally free, what it demands is mostly effort on the part of teachers and the courage to take up their challenge, a challenge that even in the short run will make teaching and learning more effective and hence more rewarding.

One of the reasons *Interdisciplinary Approaches* is such an important book is because it is structured around certain key ideas that I find essential for education generally and especially for education in the arts. At one time these ideas, which I enumerate below, were generally understood and accepted as part of the culture of art. But in recent decades, as Modern art increasingly has been characterized in purely formal terms, these ideas have been forgotten with the result that

Modernist idealism has devolved into Postmodernist irony and perhaps even cynicism (cynicism certainly would account for all the talk of late about the "end of art"). Several of the most important ideas inherent in the argument the authors put forth in *Interdisciplinary Approaches*, at least as I see them, are as follows:

> 1) Art is a serious endeavor, as serious as any other academic endeavor and should be treated as such by art and non-art teachers and administrators alike.
> 2) Art is not solipsistic; it is not something focused solely on itself as the "art for art's sake" argument would have it.
> 3) Art actually has meaning and because it has meaning one can't make art about nothing, one must make art about something.
> 4) The "something" of which art is about must be learned in order for someone to become an artist and recuperating this "something" is what is meant when we speak of learning about art, whether art of the present or the past does not matter.
> 5) Because art is a system of symbolic and iconic signs that are intended to communicate something to a viewer, art always involves community (i.e., shared values) otherwise there is no possibility for meaning to occur.
> 6) Because both the making and viewing of a work of art entails understanding communal values, to understand a work from another time or culture necessitates learning about that culture's communal values.
> 7) Art offers a unique way of understanding the world and the things and people in it that is inherently different from any understanding offered by either the social sciences or the hard sciences because the art operates through the aesthetic dimension.

What the authors are arguing, in short, is that with a work of art more is involved than just an object, event, or performance. A work of art is more than just another form of entertainment in the mode of commercial popular culture; it is not something to be looked at, watched or heard and then quickly forgotten as it is replaced by the next new enthusiasm. Within every work of art there exists something important that demands the attention of the viewer because it demands to be understood. This is why promoting, encouraging, and facilitating understandings of works of art, as the authors so cogently argue, is (or should be) among the goals of art education. But, the first thing necessary for understanding to occur, however, is for the viewer, in this case the student to accept work's challenge to be understood. Unfortunately, this is no easy task for the teacher to achieve given the nature of our society and its emphasis on consumption as a form of personal fulfillment and satisfaction.

The challenges that teachers must face when engaging students in understanding works of art bring me to what I see as another important feature of *Interdisciplinary Approaches to Teaching Art in High School.* This is its strategy; to achieve the goal of understanding by creating a dialogue between teacher and student about and around the work of art. The authors' proposal of a dialogical method is a wonderfully simple and efficient way to encourage students to accept the challenge to understanding that the work of art presents while, at the same time, de-compartmentalizing the subject matter of teaching and learning. Thus thinking and talking about art need not be relegated to the manipulation of materials and the development of technique as if they exist in a vacuum; it now can and should involve the political, social, historical, scientific, and any other subject that the work calls to mind. In this way the dialogical method not only illuminates the work of art, which easily lends itself to a network of connections and interconnections that span the aesthetic to the zoological, it helps dispel the hierarchy of learning that privileges science and its view of the world above all other views. At the same time the dialogical method encourages students to see the interrelatedness of the world and the things and people in it. A perfect example of this that is cited by the authors is artist Merle Laderman Ukeles' *Methanogenesis Project*. Done as a Mountain Lake Workshop project in 1994, it was initiated by artist and Workshop director Ray Kass and involved Kass, Ukeles, microbiologist Greg Ferry from Virginia Tech University, and students and members of the local community. The project's artistic, social, historical, and environmental dimensions began from its recreation of Alessandro Volta's 1776 scientific experiment in which he discovered methane gas in a swamp. From its beginnings as a purely scientific experiment the project went on to explore questions of the carbon cycle and plant degradation in relation to the environment and other important non-scientific issues thus opening up a dialogue around project as a work of art that involved the real human concerns of the greater community.

As Ukeles' *Methanogenesis Project* shows, engaging in a genuine dialogue, the kind proposed by the authors in *Interdisciplinary Approaches to Teaching Art in High School*, leads its participants beyond their original horizons, beyond the limitations of their prior understanding and knowledge. Because, as a process of inquiry, dialogue tends to take on a life of its own carrying along its participants with it into the unknown. Often unanticipated and unintended realms of thinking and knowing are thus opened up to participants of the dialogue as they are carried along by it, almost in spite of themselves. In this way they come to see and understand events and even each other in new and deeper ways that could not have been predicted. In the process, dialogue opens its participants to a new degree of self-understanding as well--after all, the self is always understood in relation to other selves, never in isolation; knowing the other better helps us know ourselves better.

Such dialogue should be natural to human beings, as natural as the old telling stories to the young so that memories and experiences about the past become vehicles for learning about the present. Unfortunately, today this process has been short-circuited in our society as the revolt against the authority of tradition that characterized Modernism's transformative intent has become more rhetorical than substantive. Creating a genuine dialogue in such circumstances is very difficult; it demands special techniques and skills and even patience on the part of the teacher. In their understanding and insight into these issues and problems, the authors of *Interdisciplinary Approaches to Teaching Art in High School* offer a way to make this happen that has every possibility of success.

Dr. Howard Risatti is a Professor Emeritus of Modern and Contemporary Art and Critical Theory in the Department of Art History at Virginia Commonwealth University. Before receiving his Ph.D. in art history from the University of Illinois, he earned BA and MA Degrees in music theory and composition and is ABD in theory and composition. His writings on art and craft have appeared in various journals including the Art Journal, Artforum, New Art Examiner, Artscribe, Latin American Art, The Chronicle of Higher Education, The Woman's Art Journal, Art Criticism, The Studio Potter, and Ceramics: Art & Perception. Most recently he wrote on Michael Heizer's Double Negative for Sculpture Magazine and on "Contemporary American Ceramic Trends" for the Korean publication Ceramics Monthly. In Fall 2003, he spoke on functional crafts at the Cheongju Craft Biennial in Korea and on Jackie Matisse's CAVE/virtual reality art at the "Art and New Technologies" conference in Chalon, France. He has written numerous essays for exhibition catalogues. His first book was New Music Vocabulary and appeared in 1975 (University of Illinois Press); Postmodern Perspectives: Issues in Contemporary Art appeared in 1990 and the 2nd edition in 1998 (Prentice Hall). The Mountain Lake Workshops: Artists in Locale (1996, Anderson Gallery & VA Tech Foundation) accompanied the exhibition of the same title. His latest book, Skilled Work: American Craft in the Renwick Gallery (Smithsonian Institution Press), was co-authored with Kenneth Trapp in 1998. Most recently Risatti has curated an exhibition on art in craft media for The Zone; Chelsea Gallery in New York City and has written on Craft and Design for MetalSmith Magazine and on Meaning in Craft for Ceramics Monthly.